Design Thinking for Engineers

Dr. Devaraj E
Dr. Chandrashekhar Bendigeri

CWP

This edition has been published by Central West Publishing PTY LTD,
(ABN 13 683 898 722) Australia

For more information about the books published by Central West Publishing PTY LTD, please visit https://centralwestpublishing.com

Disclaimer

Every effort has been made by the publisher, editors and authors while preparing this book, however, no warranties are made regarding the accuracy and completeness of the content. The publisher, editors and authors disclaim without any limitation all warranties as well as any implied warranties about sales, along with fitness of the content for a particular purpose. Citation of any website and other information sources does not mean any endorsement from the publisher, editors and authors. For ascertaining the suitability of the contents contained herein for a particular lab or commercial use, consultation with the subject expert is needed. In addition, while using the information and methods contained herein, the practitioners and researchers need to be mindful for their own safety, along with the safety of others, including the professional parties and premises for whom they have professional responsibility. To the fullest extent of law, the publisher, editors and authors are not liable in all circumstances (special, incidental, and consequential) for any injury and/or damage to persons and property, along with any potential loss of profit and other commercial damages due to the use of any methods, products, guidelines, procedures contained in the material herein.

NATIONAL LIBRARY OF AUSTRALIA

A catalogue record for this book is available from the National Library of Australia

ISBN (print): 978-1-922617-76-7

Foreword

In a world characterized by rapid change, uncertainty, and complexity, the ability to innovate and adapt is essential for organizations to thrive. Design thinking has emerged as a powerful approach to problem-solving and innovation, offering a human-centered and collaborative framework for addressing complex challenges and creating meaningful solutions.

This comprehensive guide to design thinking explores the principles, processes, and tools that underpin this transformative approach to innovation. Divided into five modules, the book provides a holistic view of design thinking and its applications across various domains and disciplines.

The evolution of engineering education demands an integration of analytical skills with creative and empathetic problem-solving abilities. The Design Thinking course modules presented herein are a thoughtful response to this call. Each module is designed to progressively guide students through the design thinking process empowering them to approach problems not just with logic, but with insight and innovation.

This modular structure reflects our commitment to experiential learning and interdisciplinary collaboration. From understanding human behavior to articulating final presentations, each segment of the course is carefully sequenced to promote depth, reflection, and real-world application.

It is with this vision that we present the module-wise breakdown a roadmap to cultivating design sensibilities, critical thinking, and purpose-driven innovation among our students.

In essence, this book serves as a comprehensive guide to design thinking, offering insights, tools, and techniques that can inspire creativity, foster collaboration, and drive meaningful change. Whether you are a seasoned innovator or a newcomer to the world of design thinking, this book provides a roadmap for unlocking your creative potential and driving innovation in your organization. We invite you to embark on this journey of exploration and discovery, as we uncover the transformative power of design thinking in driving innovation and shaping the future of organizations.

Preface

Welcome to the world of design thinking—a dynamic and transformative approach to problem-solving and innovation. As we stand on the cusp of a new era defined by rapid change, uncertainty, and complexity, the need for innovative solutions has never been more pressing. In this preface, we would like to offer a brief overview of what you can expect from this book and why design thinking is such a powerful tool for addressing the challenges of our time.

Design thinking is not just a methodology; it's a mindset—a way of approaching problems with empathy, creativity, and collaboration. At its core, design thinking is about understanding the needs of users, exploring new possibilities, and iteratively refining solutions based on feedback and experimentation. Whether you're designing a product, service, or experience, design thinking provides a structured yet flexible framework for driving innovation and creating value.

In this book, we will explore the principles, processes, and tools that underpin design thinking, delving into its applications across various domains and disciplines. Throughout the book, you will encounter real-world case studies, and hands-on exercises designed to deepen your understanding of design thinking and inspire you to apply its principles in your own work. Whether you're a business leader, entrepreneur, designer, or educator, this book is for anyone interested in driving innovation and creating positive change in the world.

As we embark on this journey together, we encourage you to approach the material with an open mind and a spirit of curiosity. Design thinking is not just a set of tools; it's a way of thinking and problem-solving that can transform how we approach challenges and unlock new opportunities for growth and innovation.

We hope that this book serves as a valuable resource and guide on your journey into the world of design thinking. May it inspire you to embrace creativity, collaboration, and empathy as you work to create a better future for yourself and those around you?

Pedagogical Approach for Design Thinking Textbook

In designing the pedagogical approach for this design thinking textbook, our aim is to create an engaging, interactive, and effective learning experience that empowers readers to develop a deep understanding of design thinking principles and practices. The pedagogical approach encompasses the strategies, methodologies, and principles used to facilitate learning and promote active engagement among readers. Here's how we'll approach it:

Experiential Learning: We believe in learning by doing. The textbook will incorporate experiential learning activities, case studies, and real-world examples to allow readers to actively engage with the material and apply design thinking concepts in practical scenarios. By participating in hands-on exercises and projects, readers will deepen their understanding and develop practical skills.

Collaborative Learning: Collaboration is essential for fostering creativity and innovation. The textbook will encourage collaborative learning through group activities, discussions, and peer feedback. Readers will have opportunities to collaborate with classmates, colleagues, or online communities to share ideas, perspectives, and insights.

Multimodal Content Delivery: People learn in different ways, so the textbook will deliver content through a variety of modalities, including text, images, videos, interactive simulations, and multimedia resources. This multimodal approach accommodates diverse learning styles and preferences, ensuring accessibility and engagement for all readers.

Scaffold Learning: Design thinking is a complex process that involves multiple stages and skills. The textbook will scaffold learning by breaking down concepts into manageable chunks, providing clear explanations, and offering step-by-step guidance for applying design thinking methods. Each chapter will build on previous knowledge, gradually increasing in complexity and depth.

Contextualized Learning: Design thinking is most effective when applied to real-world problems and contexts. The textbook will contextualize learning by presenting case studies, examples, and scenarios from a variety of industries and disciplines. Readers will learn how design thinking principles can be applied in diverse contexts, from product design to social innovation.

Feedback and Reflection: Feedback and reflection are essential for learning and growth. The textbook will incorporate opportunities for self-assessment, peer feedback, and reflection at key points throughout the learning process. Readers will have opportunities to evaluate their progress, identify areas for improvement, and reflect on their learning journey.

Inclusive Design: Design thinking is inherently inclusive, and the textbook will reflect this value by ensuring accessibility and representation for all readers. The content will be designed with diverse learners in mind, considering factors such as language, culture, ability, and background. Universal design principles will guide the development of content and resources to accommodate the needs of all readers.

By embracing these principles of experiential learning, collaboration, multimodal content delivery, scaffold learning, contextualization, feedback and reflection, and inclusive design, this design thinking textbook aims to empower readers to become proficient practitioners of design thinking and drive innovation in their personal and professional endeavors.

Acknowledgments

The creation of this textbook is a collaborative effort, and we extend our sincere gratitude to all those who have contributed to its development. Special thanks to the faculty members, industry experts, and students whose insights have enriched the content.

Writing a comprehensive textbook on design thinking is a collaborative effort that involves the contributions and support of numerous individuals and organizations. We would like to extend our sincere gratitude to everyone who has played a role in bringing this project to fruition.

A special thanks to the educators, practitioners, and experts who generously shared their knowledge, experiences, and feedback. Their input and insights have helped ensure the relevance, accuracy, and practicality of the content, making this textbook a valuable resource for both students and professionals.

We are grateful to the reviewers and beta readers who provided invaluable feedback and suggestions during the development process. Their careful scrutiny and constructive criticism have helped refine the content and improve its clarity, organization, and effectiveness.

We would also like to thank CMR University that have supported to complete the book, whether through funding, resources, or collaboration. Their support has been essential in making this textbook accessible and impactful for a wide range of learners.

Last but not least, we want to express my heartfelt appreciation to the readers and users of this textbook. Your curiosity, engagement, and commitment to learning are the ultimate driving force behind this endeavor. It is my sincere hope that this textbook will inspire you to embrace the principles of design thinking and empower you to tackle complex challenges with creativity, empathy, and innovation.

Thank you to everyone who has contributed to this project. Your support, expertise, and enthusiasm have been truly invaluable, and we are honored to have been part of this journey with you.

Table of Contents

Module 1

Exploring Needs and Behaviors

1.1 Introduction to Design Thinking

"A concept that sells" is something that everyone appreciates. Unfortunately, this is an outcome rather than a method for accomplishing the aim. How can one increase the likelihood that the investments in research, development, and marketing will result in a successful innovation? What are designers and engineers truly thinking and doing when they build goods, services, and businesses has been our overarching question for the past few years. Using the findings from our research as a foundation, we have created new tools, tasks, and principles that strengthen team, individual, and organizational capacity for design innovation.

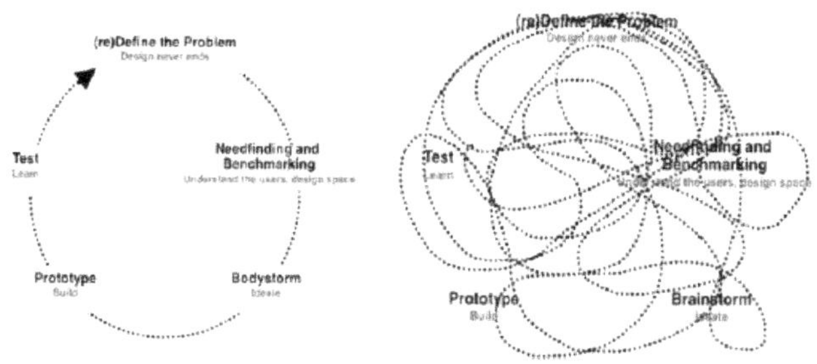

Figure 1.1 Design thinking is commonly visualized as an iterative series of five major stages

Its human-centered approach incorporates knowledge from the design, social sciences, engineering, and business fields. It combines a user-centered approach with cross-disciplinary cooperation and iterative improvement to create ground-breaking products, systems, and services. Through quick conceptual prototyping, design thinking produces a lively, dynamic environment that encourages learning. The construction and adaptive application of a set of behaviors and values are key components of design thinking. This objective is in direct opposition to, though complementary to, the predominate disciplinary approach that is centered on the development and approval of a body of knowledge.

The majority of explorations are nonlinear and fluid, or at least they should be. A true exploration is not a hurried march from Point A to Point B, but a winding path that leads to Point B with the freedom to stop along the way to take in the scenery and, perhaps, come across something new or previously unnoticed. The convoluted nature of design thinking is not due to a designer's lack of organizational and purposeful discipline. Contrary to popular belief, design thinking deliberately aims to be convoluted and fluid in order to challenge traditional approaches to issue solving.

Design thinking is a non-linear, iterative process that teams use to understand users, challenge assumptions, redefine problems and create innovative solutions to prototype and test. Involving five phases - Empathize, Define, Ideate, Prototype and Test - it is most useful to tackle problems that are not well-defined or unknown.

When attempting to solve an issue, we are skilled at asking "What's next?" (And some could even say that we have been trained in this). (I attribute this to algebra, which must stick rigidly to a sequential approach to problem-solving.) However, there isn't always a clear "next" in design thinking that we should take. Each design thinking phase may result in a number of potential "nexts", and it is up to the design thinking team to choose which "next" to pursue. It's true that initially, this could seem a little perplexing. Figure 1.2 shows the steps involved in Design thinking.

Figure 1.2 Steps involved in design thinking

1.2 Importance of Design Thinking

"Everyone appreciates an invention, a good concept." There aren't many shorter definitions of innovation. It is quite incisive. But in doing so, it exposes the fact that selling depends on elements that go outside the bounds of innovation. The directive "let's get creative" does not choose its own course. How many different ways can we define innovation, to put it another way? The question of how many different ways the inventive firm can be organized is a consequence. How many different ways can the innovation process be set up for a third iteration? We now have a concern that has to be addressed. In order to find breakthroughs that people will buy, the "Understanding Innovation" book series exposes readers to the breadth and depth of design thinking methods. In fact, it would be a mistake to think that there is only "one proper response." Instead, we present a radical, current, and rigorous voyage of discovery.

The capability to recognize and adapt to rapid changes in users' environments and actions is essential for user experience (UX) design. Since Herbert A. Simon, a cognitive scientist and Nobel Prize winner, first mentioned design thinking in his 1969 book The Sciences of the Artificial and later made significant contributions to its principles, the world has grown more linked and complex. In order to fulfil human needs in the modern era, experts from a number of fields—including engineering and architecture—then advanced this extremely creative process.

Design thinking is a useful tool that business in the twenty first century from a variety of industries use to solve problems for the consumers of their goods and services. Design teams employ design thinking to address well-defined/unknown difficulties (often referred to as "wicked problems") because they can reframe these issues in terms of the needs of users. Design thinking is unquestionably the best methodology for "thinking beyond the box" among all design methodologies. With it, organizations may conduct better usability testing, prototyping, and UX research to find fresh methods to satisfy users' needs.

Design thinking is valued as a business strategy that may improve the world (global companies such as Apple, and Airbnb have used it well). This worth is reflected in how well-liked it is as a course at prestigious worldwide universities. Teams have the freedom to come up with novel solutions while using design thinking. With its aid, your team can uncover obscure insights and employ a variety of practical techniques to aid in the discovery of novel solutions.

1.3 The Five Stages of Design Thinking

A brief or demand are transformed into a final item or design solution through the process of design. The five stages of the design process are empathized, define, ideation, prototype, and test. Design thinking is required for each of these. Following this chapter, succeeding chapters will examine particular stages of the process in more detail. This chapter will cover each of the seven stages and the design thinking features they contain. High levels of creativity are utilised during the design process, but in a way that is controlled and guided by the process to provide a realistic, workable solution to the design problem that meets or exceeds the stated objectives of the brief. Design is an activity that serves both economic and creative aims, even though creativity in design is crucial. The design process aids in ensuring that a design satisfies each of these requirements. The approach aims to develop a variety of potential solutions and makes use of several tools or mechanisms that inspire participants to consider novel or creative alternatives.

A design methodology called "Design Thinking" offers a problem-based approach to problem solving. By understanding the human requirements at play, re-framing the issue in human-centric ways, coming up with a lot of ideas during brainstorming sessions, and taking a hands-on approach during prototyping and testing, it is incredibly helpful in handling complicated challenges that are poorly defined or unknown. Anyone may use the Design Thinking approaches to address complex problems that arise in our day-to-day lives in our businesses, countries, and the rest of the world by having a solid understanding of these five stages of the process.

The concentrate on the Stanford Hasso-Plattner Institute of Design's five-stage Design Thinking approach (d.school). The top university for teaching Design Thinking is the d.school. According to d.school, the five steps of design thinking are empathize, define (the issue), ideate, prototype, and test. Let's examine each of the five stages of design thinking in more detail.

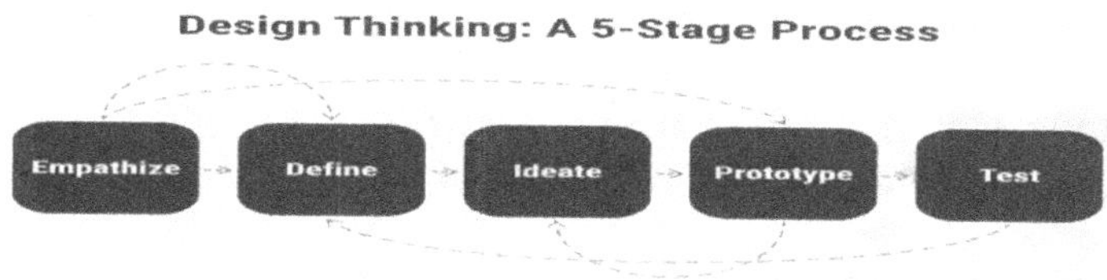

Figure 1.3 Design Thinking Stages

Stage 1: Empathize - Research Your Users' Needs

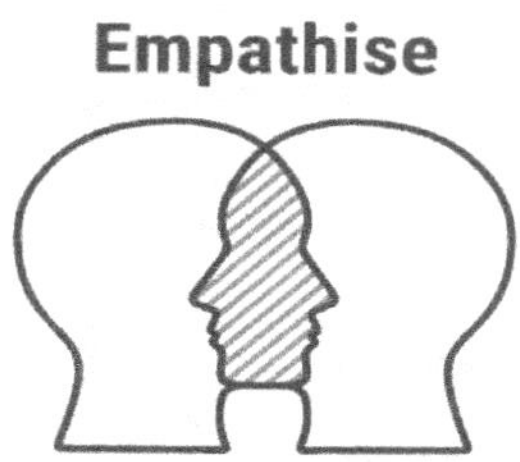

Figure 1.4 Empathize

The key is to put oneself in the user's or customer's shoes and pay close attention to him. Empathy can put the inventor in a position where they are both far from themselves and close to their customers at the same time. To put it another way, this strategy fosters client orientation. Thus, improvements can be more closely connected with the needs of the customers and, if necessary, prioritized according to how much they can fulfil those customers' wants and needs. Empathize in general means to look the problem in other person point of view.

Developing an empathic knowledge of the issue you're seeking to address is the first step in the Design Thinking approach. By observing, interacting with, and empathizing with people to understand their experiences and motivations, as well as by immersing yourself in the physical environment so you can gain a deeper personal understanding of the issues involved, you can consult experts to learn more about the area of concern. A human-centered design method like Design Thinking depends on empathy, which enables design thinkers to put aside their personal worldviews to better understand consumers and their needs.

Depending on the time restrictions, a significant quantity of data is acquired at this stage to be used in the following stage and to create the best understanding of the users, their demands, and the issues that drove the creation of that specific product.

Stage 2: Define - State Your Users Needs' and Problems

Figure 1.5 Define

This stage involves thorough investigation into the requirement or issue raised by the client as well as on-site observations. This can be accomplished using a variety of techniques, including interviews, written surveys, observations with recording through images, or even movies. As a result, the general circumstances have been clarified, the target group has been precisely defined, and a thorough grasp of the customer's demands and behavior has been achieved.

During the Define stage, you put together the information you have created and gathered during the Empathize stage. This is where you will analyze your observations and synthesize them to define the core problems that you and your team have identified up to this point. You should seek to define the problem as a problem statement in a human-centered manner.

For instance, a much better way to explain the problem would be, "Students need to eat nutritious food in order to be healthy," rather than as your own want or a need of the company, such as, "We need to raise our food-product market share among by students by 10%."

The Define stage will assist the designers on your team in generating excellent ideas for features, functionalities, and any other components that will enable them to either solve the problems or, at the at least, make it as simple as possible for users to solve problems on their own. By asking questions that can assist you look for ideas for solutions, you can get from the Define stage to the Ideate stage more quickly. For example, you might ask: "How might we... inspire students to perform an action that benefits them and also incorporates your company's food-product or service?"

Stage 3: Ideate - Challenge Assumptions and Create Ideas

Figure 1.6 Ideate

Designers are prepared to begin coming up with ideas at the third stage of the Design Thinking process. In the Empathize stage, you developed an understanding of your users and their needs. Then, in the Define step, you analyzed and synthesized your observations to produce a human-centered issue statement. With this strong foundation, you and your team may begin to "think outside the box" to develop novel solutions to the problem statement you have created, as well as start searching for different perspectives on the issue. There are countless methods for coming up with ideas, including Brainstorming, Brain Writing, Worst Possible Idea, and SCAMPER. It is common practice to use brainstorming and "worst possible idea" sessions to encourage creative problem-solving. The Ideation phase should start with as many ideas or solutions to the problem as possible. By the end of the Ideation phase, you should choose a few additional Ideation techniques to assist you in exploring and testing your ideas and determining the most effective means of either resolving a problem or supplying the pieces necessary to get around it.

Stage 4: Prototype - Create Solutions

The design team will now create a variety of reasonably priced, scaled-down copies of the product or particular product aspects so they may research the problem solutions developed in the earlier stage. The design team itself, other departments, or a small number of individuals outside the design team may exchange and test prototypes. In this experimental stage, which follows the first three, the goal is to find the best solution for each of the issues that were discovered.

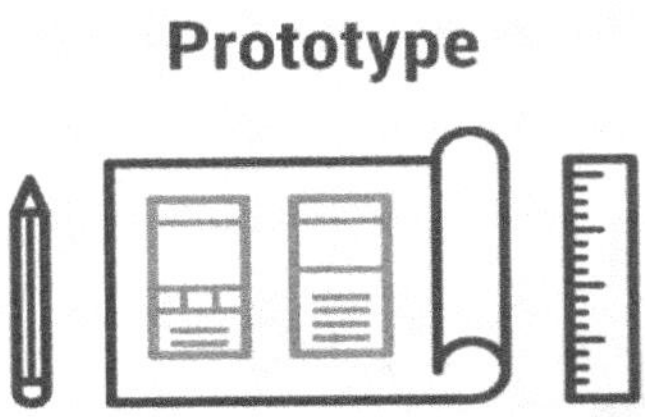

Figure 1.7 Prototype

The answers are put into practice in the prototypes, and each one is analyzed and either accepted, improved, and reexamined, or rejected based on the experiences of the users. By the end of this phase, the design team will have a better understanding of the limitations of the product and the issues that exist, as well as a clearer understanding of how actual users will act, think, and feel while interacting with the finished product.

Stage 5: Test - Implement Your Solution

Figure 1.8 Test

The best options found during the prototyping phase are extensively tested by designers or assessors on the entire product. This is the last stage of the five-stage model, however in an iterative process, the results from the testing phase are frequently utilized to redefine one or more problems and inform knowledge of the users, the usage context, human thought, behavior, and emotion, as well as the ability to empathize. To rule out potential solutions to problems and gain the most comprehensive understanding of the product and its users feasible, changes and adjustments are still done even at this phase.

1.4 The Non-Linear Nature of Design Thinking

We may have described a straightforward and linear Design Thinking approach, where each stage appears to lead logically into the next and end with user testing. The process is carried out in a more flexible and non-linear manner though. For instance, separate teams within the design team might work on many stages at once, or the designers might gather data and create prototypes throughout the entire project to help them visualize the problems and bring their ideas to life. A new brainstorming session (ideate) or the creation of new prototypes may arise from the testing phase's results, which may also provide some user insights (Prototype).

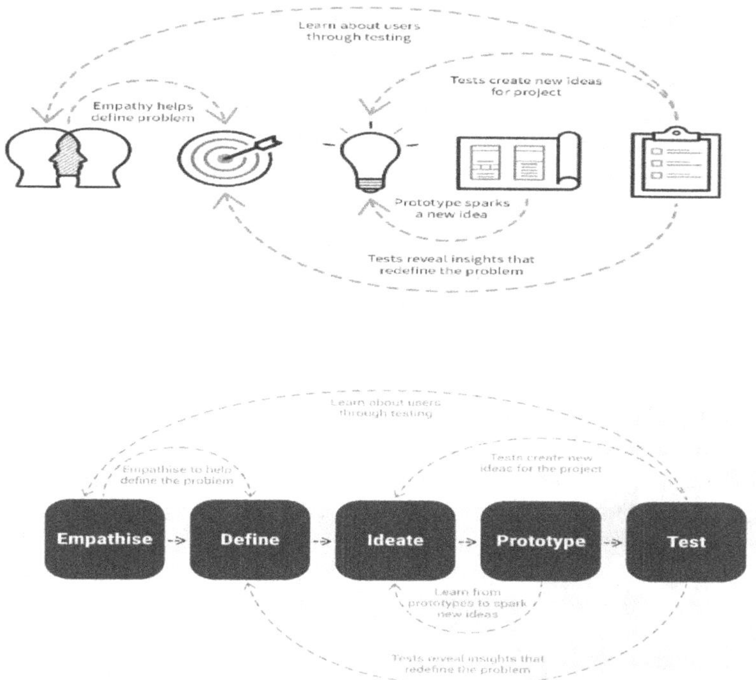

Figure 1.9 Design thinking: A non-linear process

It is important to understand that the five stages are not always sequential. They can frequently occur simultaneously and be repeated iteratively, and they are not needed to occur in any order. As a result, rather than being seen as a series of processes, the stages should be viewed as various ways that contribute to a project. The five-stage Design Thinking approach, on the other hand, systematizes and specifies the five steps you would anticipate completing in a design project, as well as in any innovative problem-solving attempt. Every project will have tasks unique to the product being developed, but each stage's core concept will always be the same.

Design Thinking should not be viewed as a rigid and immovable method to design; rather, the steps it consists of, as shown in the above example, serve as a general framework for the actions you would usually do. These steps may be switched, carried out simultaneously, and repeated numerous times in order to broaden the solution space and focus in on the best options for your specific project. This will help you acquire the clearest and most insightful ideas.

One of the key advantages of the five-stage approach is the way that knowledge gathered at later phases may feedback to early stages, as you can see from the example above. Information is continuously used to reframe the problem and to improve understanding of the problem and solution domains (s). As a result, a never-ending cycle is created in which the designers continuously learn new things, establish fresh perspectives on the product and its potential applications, and obtain a deeper understanding of the users and the issues they encounter.

The Origin of the 5-Stage Model

Nobel Prize winner Herbert Simon presented one of the first official models of the Design Thinking process in his key book on design methodologies, "The Sciences of the Artificial," published in 1969. The seven key phases of Simon's model, each with its own component stages and activities, had a significant impact on the development of some of the most popular versions of the Design Thinking process in use today. Although there are numerous variations of the Design Thinking method in use today, with stages varying from three to seven, they are all based on the same concepts found in Simon's

1969 model. Our primary focus is the Stanford Hasso-Plattner Institute of Design's five-stage Design Thinking approach (d.school).

1.5 Analysis of Secondary Research

Secondary research, commonly referred to as desk research, is a type of inquiry that entails gathering material that has already been obtained through a number of sources. This refers to both internal sources (such as internal research) and, more frequently, external sources (such as government statistics, organizational bodies, and the internet). Secondary research can be sourced through websites, libraries, and museums and comes in a variety of forms, including published datasets, reports, and survey results.

The data is typically free or available for a small fee and is acquired through surveys, phone interviews, observation, face-to-face interviews, and other methods. Researchers gather, verify, analyze, and incorporate secondary research to support the confirmation of their research objectives for the study period. In addition to the aforementioned, it can be used to review earlier studies in a certain field. Researchers can use it to verify initial hypotheses and determine whether it's worthwhile to continue study into a potential field by looking for patterns across data spanning multiple years and identifying trends.

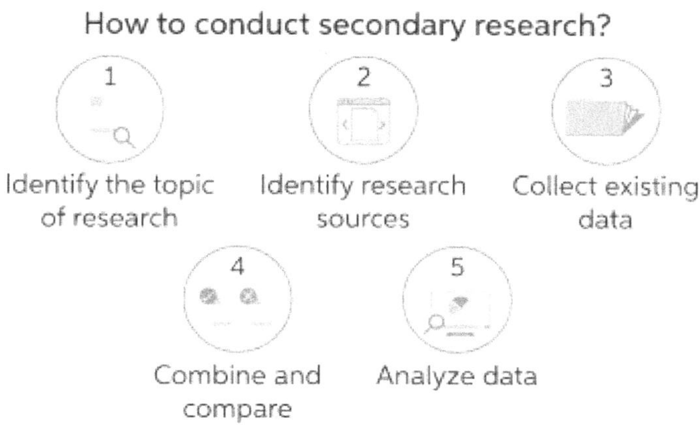

Figure 1.10 How to conduct secondary research?

How to Conduct Secondary Research?

There are five key steps to conducting secondary research effectively and efficiently:

1. Identify and define the research topic

Understand the subject of your research before defining it by considering the research questions you hope to have resolved.

What is the purpose of undertaking this research, you could ask? Then, consider: What do we hope to accomplish? This might suggest a possible explanation for why something occurred or support a theory. The responses may point to concepts that require primary or secondary research (or a combination of the two) to be investigated.

2. Find research and existing data sources

If secondary research is required, consider potential sources of the data. This aids in focusing your secondary sources on those that address the queries you have. What keywords are necessary?

Which organizations are already actively engaged in this work? Do you need to be aware of any competitors?
Make a list of the information, data, and individuals who can assist you with your assignment.

3. Begin searching and collecting the existing data

Start gaining access to the data and compiling it into a system once you have the list of data sources. This could entail opening up research journal accounts or dialing third-party research teams to schedule meetings to confirm the specifics surrounding data outcomes.

Check the data's date, the reliability of the source, the material's applicability to your study topic, and the research process employed by the outside researchers as you look for and access information. Start small, and when you see progress, expand your research into the areas that will most benefit your study's objectives.

4. Combine the data and compare the results

You must intelligently comprehend, filter, order, and merge your data once it is all in one location. Data may come in a variety of formats, some of which may not be useable, necessitating the deletion of other data.

After that, you may start examining various data sets to see what they have to say. You can discover that in order to detect overlaps or patterns, you need to compare the same datasets over other time periods or distinct datasets. What does this data represent for my research, you ask? Does it aid or interfere with my research?

5. Analyze your data and explore further

Look at the data you have at this point in the procedure and consider whether it addresses the initial research questions you had in mind. Do you notice any gaps? Do you comprehend the data you've discovered? Repeat the process and go deeper into the subject if you think there is more to learn so you have all the information you require.

Consider including primary research data to your findings to complete your analysis if secondary research is unable to provide these answers. As you continue to research and learn more, update your findings. This will help you deliver accurate, trustworthy information.

A study technique called secondary research, sometimes known as desk research, uses data that has previously been collected. To improve the overall effectiveness of research, existing data is compiled and evaluated.

Research that has already been published in research reports and other comparable papers is considered secondary research. These materials may be made available through online resources, public libraries, surveys that have previously been completed, etc. Some governmental and non-governmental organizations also keep data that can be obtained and used for research. It is far more affordable than primary research since it uses data that is already available, as opposed to primary research, which collects data directly from organizations or businesses or has them hire a third party to do so.

Secondary Research Methods with Examples

One of the things that makes secondary research a popular option for many firms and organisations is how affordable it is. Not every company has the financial means to spend a lot of money on research and data collection. Since material may be retrieved while seated at a desk, secondary research is also appropriately referred to as "desk research".

Following are popularly used secondary research methods and examples:

1. Data available on the internet: Internet use is one of the most often used methods for gathering secondary data. On the internet, information is easily accessible and can be downloaded with the press of a mouse.

One may have to pay a little fee to get already-existing data, but this data is essentially free. Websites have a wealth of data that businesses and organizations can use for their research requirements. To collect information, organizations must, however, only use reliable and trustworthy websites and also the quantity of collected data.

2. Government and nongovernment agencies: Some government and non-government organizations may also provide data for secondary research. For instance, businesses and organizations can use the useful and pertinent data provided by the Government Printing Office, the Census Bureau, and Small Business Development Centers.

To download or use the data offered by these agencies, there may be a fee. These organizations provide reliable and authentic data.

3. Public libraries: Another excellent place to look for information for secondary research is public libraries. Copies of significant earlier research are available in public libraries. They serve as a repository for crucial data and papers from which data can be gleaned.

Each of these public libraries offers a different range of services. A significant collection of company directories, bulletins, periodicals, and government publications featuring market information are frequently found in libraries.

4. Educational Institutions: It's common to undervalue the value of gathering information from educational institutions for secondary research. In contrast to the commercial sector, however, useful research is occasionally undertaken in colleges and universities as well.

Universities mostly use the data they gather for primary research. Businesses and groups can, however, go to educational institutions and ask for information from them.

5. Commercial information sources: Local radio, TV, newspapers, journals, and magazines are excellent places to find information for secondary research. These commercial information sources provide first-hand information on topics like demographic segmentation, market research, political agenda, and economic changes.

Businesses or other groups might ask for the data that is most pertinent to their investigation. Businesses have the chance to identify their potential customers, but they can also learn about the ways to advertise their goods or services through these sources because of their greater reach.

Key differences between primary research and secondary research:

Primary Research	Secondary Research
Research is conducted first hand to obtain data. Researcher "owns" the data collected.	Research is based on data collected from previous researches.
Primary research is based on raw data.	Secondary research is based on tried and tested data which is previously analyzed and filtered.
The data collected fits the needs of a researcher, it is customized. Data is collected based on the absolute needs of organizations or businesses.	Data may or may not be according to the requirement of a researcher.
Researcher is deeply involved in research to collect data in primary research.	As opposed to primary research, secondary research is fast and easy. It aims at gaining a broader understanding of subject matter.
Primary research is an expensive process and consumes a lot of time to collect and analyze data.	Secondary research is a quick process as data is already available. Researcher should know where to explore to get most appropriate data.

Advantages of Secondary Research

1. Most information is secondary research is readily available. There are many sources from which relevant data can be collected and used, unlike primary research, where data needs to collect from scratch.

2. This is a less expensive and less time-consuming process as data required is easily available and doesn't cost much if extracted from authentic sources. A minimum expenditure is associated to obtain data.

3. The data that is collected through secondary research, gives organizations or businesses an idea about the effectiveness of primary research. Hence, organizations or businesses can form a hypothesis and evaluate cost of conducting primary research.

4. Secondary research is quicker to conduct because of availability of data. Secondary research can be completed within a few weeks depending on the objective of businesses or scale of data needed.

Disadvantages of Secondary Research

1. Although data is readily available, credibility evaluation must be performed to understand the authenticity of the information available.

2. Not all secondary data resources offer the latest reports and statistics. Even when the data is accurate, it may not be updated enough to accommodate recent timelines.

3. Secondary research derives its conclusion from collective primary research data. The success of your research will depend, to a greater extent, on the quality of research already conducted by primary research.

1.6 Expert Interactions

You must pay great attention to the style of your reviews if you want to employ expert opinions to support your user research. Why? The main reason for this is because even when someone is an expert, they could still overlook something because we're all human and there is a tendency to assume that skill in one field automatically translates to expertise in other areas. Thus, the finest expert review procedure gathers a team of specialists and enables them to collaborate on the review. Let's examine a procedure that could assist you in doing that:

The Process of Group Expert Review (Part 1)

Please don't miss the second stage of this review process, which will be published on the UX Daily tomorrow and will affect how effective your work is:

Choose a Group

The ideal team will include both usability specialists and subject matter experts. Depending on the size of the project, there could be as few as 3 or as many as 12 individuals involved. If you use more than 12, you run the risk of losing people's contributions in the mix; if you use fewer than 3, you return to using too little input to conduct a complete assessment.

Figure 1.11 Expert Group interactions

Additionally, try to vary the reviewers' levels of experience. When compared to someone with enormous levels of experience, someone with limited acquaintance with the industry will frequently identify a variety of different concerns.

Customer research has typically been a cold endeavor. An expert evaluates comments from focus groups, surveys, and, if available, statistics on current behavior and makes assumptions about needs based on his or her prior notions about customer preferences. The inferences are more accurate the better the data is. The problem is that this grounds people in the requirements that have already been expressed and which the data represents. They interpret the data in light of their own prejudices. Additionally, they fail to notice unmet needs.

1.7 User-centered Research Methods

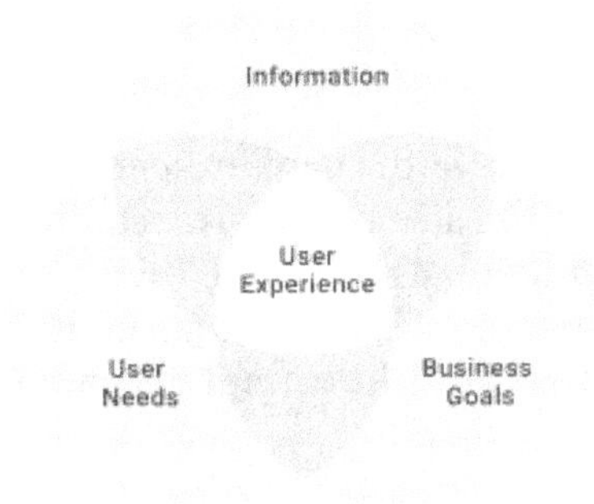

Figure 1.12 User-centered research methods

User-centered design (UCD) is an iterative design process in which designers focus on the users and their needs in each phase of the design process. In UCD, design teams involve users throughout the design process via a variety of research and design techniques, to create highly usable and accessible products for them.

The significance of user-centered design was initially discussed by cognitive scientist Dr. Donald Norman. According to him, design choices ought to be based on the requirements and preferences of consumers. In line with user-centered design's core values, there are:

- Empathy,
- Optimism,
- Iteration,
- Creative confidence,
- Belief in making,
- Embracing ambiguity,
- Learning from failure.

A special opportunity to design with communities is created through user-centered design. User-centered designers have a thorough understanding of the people they are aiming to help. They generate a tonne of ideas and develop ground-breaking new goods that are based on the needs of real people.

Identify the users of the product, the purposes for which they will use it, and the environments in which they will use it to design with the user in mind. Be smart about your challenge by learning about people's lives and their ambitions and wants.

A universal language between scientists, stakeholders, and end users is provided by user-centered design. There are six stages in the user-centered design process:

- Specify the use context and user's needs
- Specify business requirements
- Build design solutions from rough concept to finished design
- Evaluate designs with usability testing
- Implementation phase involving of the development and deliver the product
- Deployment phase involving the final product is evaluated, as consumer needs change.

1.8 Methods of User-Centered Design Research

The uniqueness of UCD is that it prioritises experimentation, iteration, failure-based learning, and research above all else. For instance, user research focuses on speaking with people about their difficulties, objectives, and constraints. But there will be times when you require more background information, background context, or data than a casual interview can provide.

Speak with the target audience directly by doing individual, group, or expert interviews. There is no better method to understand their objectives, wants, and hopes. Take notes on the comments you receive and take use of the opportunity to ask further questions and develop your ideas.

Responsive principles of user-centered design:

- Clear understanding of the users, tasks and environments;
- Evaluation-driven design;
- considering the general consumer experience;
- involving the client within the design process;

All phases of this process can employ user-centered design. The requirement clarification process and the steps the team should take to better clarify the requirements are both covered by the Interaction Design Foundation. This is how that procedure appears:

- Understand the context of its use
- Specify user requirements
- Design a solution
- Evaluate against requirements

Let's go over several methods that the product team can employ to provide a better customer experience:

Numerous research techniques are accessible, but five crucial research techniques should be emphasized based on our experience developing a user-centered financial solution. These techniques offer the finest means of obtaining key user information.

1. User Personas

It's critical to comprehend the end customers in order to develop a solution that is user-centered. It is only feasible to design the ideal product for them when we are aware of their needs, behavioral patterns, and pain points.

User personas are fictional individuals that are precisely described in terms of their age, gender, occupation, location, income, day-to-day activities, thoughts, and other characteristics. Polls, stakeholder interviews, desk research, and field research can all be used to establish the precise end-user types that each user persona represents.

One mobile banking software, for instance, features user profiles for a young student with a part-time job, a mother of two who manages her family, and a businessman who has to check both personal and professional money.

The primary purposes of this mobile banking software and how each sort of user wants to utilise it vary. By using user personas, it is feasible to develop an empathy map that depicts how the end users feel and think, the challenges they encounter on a daily basis, and their top objectives for their ideal mobile banking app.

2. User Interview

In order to understand consumers' motives, pain points, feelings, daily routines, and behaviour when using various financial products, user interviews are one of the frequently utilised research methodologies. It can be carried out at any point in the creation of a product.

User interviews will help determine whether the impending product will be used as intended in the early phases, prior to the product launch. Understanding the problems users have and the reasons why they use or don't utilize a particular feature will be useful for an existing product.

For instance, a retail bank wanted to double the volume of payments made using its mobile app. Many consumers admitted during user interviews that they hardly often or never make payments through the smartphone app. The major justifications given were that it is simpler to do the task from a computer and that they had trouble navigating the app's payment page. This inspired the bank to consider ways to make the payment form much simpler to use, which led to a rapid change in the consumers' habits of only using computers for payments.

3. Card Sorting

A technique used to comprehend how consumers organize information is card sorting. It can be done to establish an intuitive Information Architecture for a product either developing a brand-new one or when enhancing an already existing one.

There are now two different kinds of card sorting techniques: open card sorting and closed card sorting. The use of open card sorting makes it possible to predict the number, nature, and names of the groups that users will form. Will customers group their credit card information with their accounts or segregate it into a distinct group, for instance?

On the other hand, closed card sorting aids in understanding what data will be sorted into the user's given groups. For instance, the following groups are planned to be created: Accounts, Cards, Payments, Profile, and Settings Information that must be sorted into certain groups is provided to the user. If the structure adheres to the desired structure, it is therefore possible to predict how the user would arrange it.

4. Tree Testing

A tool called tree testing is used to make sure that the information architecture designed for a product is clear to users and that they can easily utilise the service to discover what they're looking for.

When developing a new product or making modifications to one that is already in use, tree testing can be done at any step of the process. This approach will provide useful insights into whether the architecture of a newly built or existing product is clear and simple to use, as well as why users may have trouble locating a certain feature.

Users are given a long collection of categories and subcategories and requested to complete specific tasks while using the tree testing approach. For instance, a user can be prompted to add an additional account or pay for utilities in a different currency. This makes it possible to assess if the user will have no trouble performing the requested task.

5. Usability Testing

The most effective way to learn exactly how actual consumers interact with a product is through usability testing. To test different situations or a whole product, it should be carried out as frequently as possible.

Users' comprehension of flows can be tested during the wireframe stage. During the design prototype stage, check that all visual accents are being used properly and observe how current customers of the service—as well as brand-new consumers - use it.

Users are asked to complete certain tasks during usability testing, such as freezing a card that a user may suspect is lost or stolen or sending a buddy a money request for 250 rupees. It's crucial for a user to comment on each step as they are completed and to express any thoughts they have as they carry out each scenario.

It is important to pay close attention to how this person handles each scenario. Five people can be tested to acquire good insights because it will give a general notion of how well the user can handle and complete the scenario.

You must have a thorough understanding of your target market—the individuals who will use your product—in order to develop an effective character. It is crucial since if you don't comprehend your target market, you can end up giving them the incorrect solution.

1.9 Impact and Opportunities of UCD

User-centered design relies on human bias to operate. Humans, their hopes, needs, and concerns are the only things driving it. The designers Figure out what is most appealing to people. However, after the designers have decided on solutions for a target community, they must enlist technical knowledge. To do this, they should devise strategies for making these solutions profitable.

Finding the correct balance is a key component in designing effective and sustainable solutions. If you want to develop solutions that affect end users, you cannot live in abstractions. Workable, desirable, and viable solutions are required.

Testing, failing, and trying again is a constant process for user-centered designers. The methodology makes use of instruments including group action, analysis, and generative processes. By including all users during the product's design and testing phases, user-centered design (UCD) builds an understanding of users' needs. Everyone in the group is involved in the designers' reliance on user feedback. The design is kept more adaptable and successful through the iteration process.

UCD has a significant impact on how well a product is launched. Microsoft is the best example of a company that has benefited from user-centric design. For a very long time, the company was heavily focused on technology. Their user experience (UX) was evolving, which was unfavorable to their users.

But over time, the organization modified its approach to become steadily more user-centered. They implemented a genuine design development approach that put their users first. Microsoft made the decision to incorporate user-centered design into their corporate culture and code of conduct.

A human bias exists in user-centered design. The most desired things are typically found by designers by starting with people and their hopes, anxieties, and requirements. The lens through which we view our solutions, however, is just one. It is time for technical knowledge once the designers have decided what the target community's solutions will be. In addition, they must devise a means of making the proposed solution profitable. To create solutions that are successful and durable, finding balance is absolutely essential. When the goal is to get impactful solutions out into the world, you cannot live in abstractions. You have to make the solutions that are desirable, workable, and viable.

How UCD Changes the Design Process and Customer Behavior?

User-centered designers experiment, make mistakes, and try again. This approach includes analysis, interviews, creative thinking techniques, and group action tools. By including users at every stage of planning and testing, UCD, for the most part, creates an awareness of the user's desires. We rely on input from every member of the organization. The design approach is responsive thanks to iteration.

The success of the product launch is ultimately influenced by UCD. One of the best instances of how user-centered design works is Microsoft Corporation. It was a technology-focused firm for a very long time. The company has since modified its strategy to be user-centered. They have embraced a genuine design development methodology that prioritizes users.

Their user experience (UX) was getting more and more sophisticated, which was bad for the people. Microsoft made the decision that their coding should incorporate user-centered design. A unique development process and teamwork gave this software behemoth hope. Business success was a result of the user-centered design. By collaborating with end users, the business and user goals were successfully created.

1.10 Empathizing with Stakeholders

It's sometimes simpler to say than to do to identify the problem to tackle. We develop software to assist people, so in order to fully comprehend the issue, we must first comprehend the individuals whose lives will be impacted by the software we develop. The more we can relate to their requirements, the more clearly we'll be able to identify and comprehend the actual issues that require fixing.

When we have a question or worry about a stakeholder in our software, we call them. The role of the architect is to identify stakeholders and comprehend their requirements. The system's expectations from our stakeholders will either directly or indirectly affect how it is designed.

The force behind design is empathy. You will create better software if you have empathy for the users of your products. In this chapter, you'll discover how to choose whom to consult with the issue at hand and what you can learn from them to begin developing the architecture.

Talk to the Right People

Stakeholders frequently, though not necessarily, have a financial interest in the Program. They could buy the programme or directly benefit from it. Both users and those who design and run the system are important stakeholders. It is occasionally crucial to consider other people's concerns, even when they might not even be aware of how our Program might affect them.

In the wild, stakeholders hardly ever go by themselves. We use the term "stakeholder group" to underscore this idea. Working with individuals on a one-on-one basis is different from working with groups. Two people who are a part of the same stakeholder group can provide information that is inconsistent or contradictory. We must try to understand the concerns of the entire group and, on occasion, even participate in obtaining a resolution.

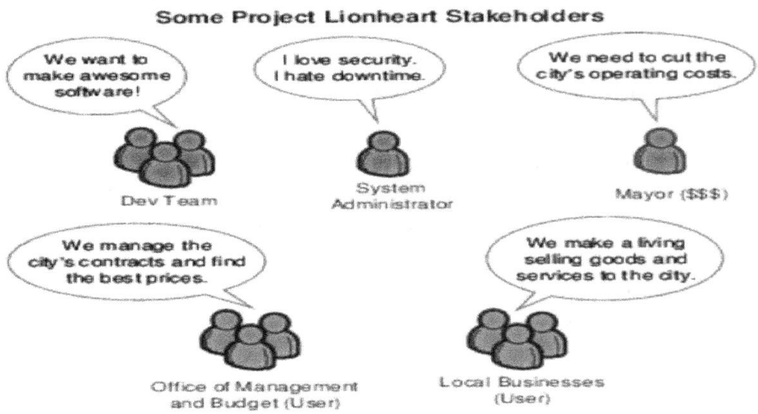

Figure 1.13 Lionheart stakeholders

The Project Lionheart Case Study's major players are listed below. In this illustration, several people or roles are represented by icons, and the speech bubbles represent the various stakeholders' feelings.

Stakeholders will affect the architecture we design and have an interest in the building we're building. Given that we'll want to invite them to upcoming design workshops, we need find out more about them. Let's move on to the map of stakeholders.

In architecture, the client comes first. If I create architecture that the consumer will not value, I will have wasted my time. When I talk to customers, I frequently overhear harrowing tales about how their current systems were designed by someone in a high-rise building who didn't understand them or their employment. How can I know that my architecture adds value for my clients?

To respond, I would use a customer-centric design technique. The first step is learning about the client's objectives. The system groups actions a user takes into categories. Find out where they struggle and how they start each assignment.

You may be thinking that this seems more like user experience than architecture. Yes, it is, but many UX designers lack the technical expertise needed to establish the architecture. My process goes above and beyond the obvious to ensure that the underlying structures uphold the client's ideals.

Step 1:

By observing how the consumer completes the task in their natural surroundings and asking several whys, you can ascertain what important to them, including their functional requirements and quality features.

Step 2:

Create a prototype of the system and document it based on the needs of the customer. The prototype shouldn't just be a flowchart; it should be as interactive as feasible.

Step 3:

As soon as feasible, go over the prototype with the client. Ensure that they are fully aware of the changes in the new system and how they will affect them.

Step 4:

Adapt the architecture in response to customer feedback.
By following these four steps, you can use your architecture to add value for your customers and avoid becoming the arrogant person in the ivory tower who is wrecking their lives.

1.11 Quantitative and Qualitative Research

The designer will have a better understanding of the examined issue thanks to the rich insights that qualitative research methodologies may offer.

Qualitative research aims to delve "beyond the surface" and is exploratory or interrogative in nature. The goal is to learn more about how people live, act, use, or require certain objects in their personal or professional lives. (Gov Design Service Manual, 2016)

A nice illustration of how qualitative research might appear is through interviews. By questioning "Why" and "How," they aid designers in understanding a person's attitudes or mentality. "It compels the subject of the interview to reflect on the causes and justifications for a certain conduct."

Through the generation of data and facts, quantitative research seeks to provide solutions to specific problems. The objective is to provide a "representation" of what consumers act or believe. 2015 (Barnham)

Surveys are an excellent illustration of quantitative research. They ask a series of specific, predetermined questions in order to determine "How much" or "How many." Following that, participants must decide which response ("Yes," "No," "Never," or "twice a week") best describes them. Off-site or indirect methods are used to perform quantitative research. It is a technique that involves many people in order to collect pertinent data. Quantitative research yields numerical or factual results. On the other hand, the lack of context makes these "facts" easily misinterpreted. Additionally, it is impossible to determine whether a participant fully understood the questions or what his underlying objective was.

Let me list the salient characteristics of both approaches: The goal of qualitative research is to comprehend the situation holistically. The researcher is searching for something, but he's not really sure what. The outcomes are insights expressed in words, phrases, or images. The data is unstructured, and the findings are the researcher's interpretation. Quantitative research, on the other hand, aims to categories, count, or measure. The researcher is already aware of the particular information he seeks. The outcomes are measurements, usually expressed as numbers. Quantitative research is organized and has an approach to prove a particular claim.

The conclusions drawn from qualitative research methodologies are always based on the researcher's interpretation and subjectivity. Despite the data' lack of reproducibility and statistical accuracy, they can nevertheless be quite insightful when used in a process for designing products with a focus on people. However, quantitative research techniques are useful at a later point of the process to validate prototypes or assess changes.

It is incorrect to oversimplify both approaches and claim that quantitative research methods are consequently objective while qualitative research methods are subjective. That the two approaches complement one another is crucially vital to understanding.

A method for practical, creative resolution of business problems. It's a form of collective solution-focused thinking with the intent of producing a constructive future result through quantitative and qualitative research.

As seen earlier, the Design Thinking Process can be synthetized in 5 key steps:

1. Emphathize
2. Define
3. Ideate
4. Prototype
5. Consumer Test

Qualitative Research is at the heart of Stages 1 and 5.

Design Thinking's initial phase is the foundation of a human-centered design approach is EMPHATHIZE. It involves comprehending the target audience for our designs as well as the major players involved in the process.

It's a hands-on exploratory component of a method for gathering information about customers that is frequently utilised in the social and behavioural sciences, in the words of qualitative researchers.

To empathize, we need to get everyone involved in the process to:

- View consumers and their behavior in the context of their lives.
- Interact with and interview consumers through both scheduled and short 'intercept' encounters.
- Experience what our consumer's experiences.

We can learn a lot about what consumers believe and feel by paying attention to what they do and how they engage with their surroundings. Understanding their needs aids us in providing for them. We can see how people's experiences, actions, and words appear physically by observing them. As a result, we will be able to comprehend the elusive significance of those experiences and gain new understanding. Your innovative solutions will be inspired by these insights. The most accurate understandings of how people behave lead to the best solutions.

We need to be able to see things "as they are," though, to acquire this new clarity. This is made possible through technologies that increase empathy and a mindset that is human-centered. This can be achieved by assigning every member of the research team to spend a day living the lives of our target consumers through conducting in-home ethnographic interviews, engaging with consumers in an online ethnographic study, or going on an observation trip to the key consumption points of our target.

After compiling all of our detailed observations and understanding about our goal, we will proceed to Stage 2 or Define.

We will share, analyse, and synthesize the field day's hundreds of findings into compelling requirements and insights, which we will then translate into issue statements, during the define stage.

This part of the process involves organizing all the data and concentrating on what's pertinent to our larger business challenge.

The essential participants in the process will meet for a half-day during which they will download all of the observations and results from the emphasis phase, synthesize them, and turn them into problem statements.

At this point, we give the workshop participants specific assignments by having them turn all of their pertinent observations and discoveries into informative post-its, one for each discovery and its source.

Then we'll ask everyone to stick their numerous Post-its to the wall so we can show the audience what they wrote on them. After being grouped into pertinent themes, these post-its will subsequently be converted into issue statements that need to be resolved.

Following the prioritization of problem statements, it is time to move on to Stage 3, also known as the Ideation stage, which is the phase of the design process where we seek to produce radical design alternatives. This often takes a half-day.

It represents a mental process of "going wide" in terms of ideas and results; it is a manner of "flaring" as opposed to "focusing." Ideation's objective is to investigate a wide range of potential solutions, including a big number of distinct ideas as well as a variety of those ideas. From this vast depository of ideas we will build prototypes to test. We ideate to move from identifying problems to finding solutions.

It's time to move on to Stage 4 of the process, Prototype, after we've ranked the top solutions in order of importance.

Creating a prototype is the process of putting concepts and experiments into the real world. A prototype can be anything that has a physical shape, including a storyboard, a wall of post-it notes, a role-playing game, a room, an item, an illustration, or an object.

The best prototypes are those that can be experienced and interacted with by people (the design team, the user, and others). The lessons we take away from those conversations can help us develop deeper empathy and create effective solutions.

Once we have solid prototypes created, it's time to test creative solutions, something we qualitative researchers perform frequently.

1.12 Human Behaviour

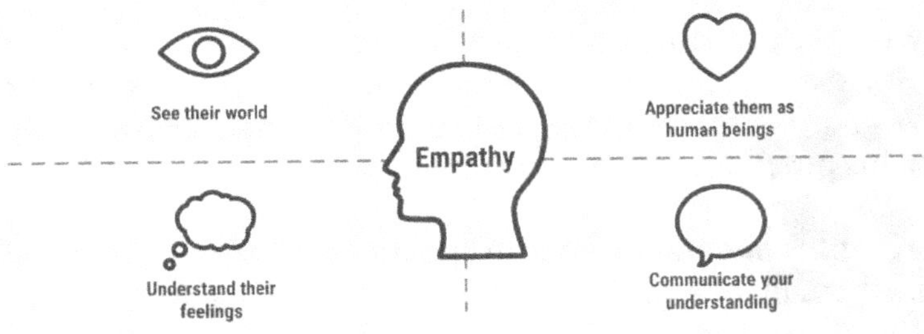

Figure 1.14 Human behavior

Important facets of imaginative engineering design are covered in "Human Behavior in Design." The connection between "picture" and "idea," two complimentary modalities, the internal and exterior elements of design thinking, and design strategies-both for lone designers and design teams-are the key issues. The objective is to enhance and assess design-supporting tools and processes.

Four quadrants that are laid out on a board, piece of paper, or table make up an empathy map and represent the four main characteristics that the users shown or possessed throughout the observation stage. The users' actions, words, thoughts, and feelings are represented by the four quadrants. While it is very simple to ascertain what users said and did, doing so requires close observation of the users' behaviour and how they reacted to various activities, ideas, chats, etc.

1.13 Design Kit: The Human-Centered Design Toolkit - IDEO

Additionally, IDEO has created contextualized toolkits that repurpose the design thinking procedure. One of these iterations concentrates on the environment for social innovation in poor nations. The vocabulary needs to be clarified for this context, made memorable, and reorganized for the normal difficulties encountered in those contexts. They created the Human-Centered Design (HCD) Toolkit for this goal, which was renamed to stand for hear, design, and deliver.

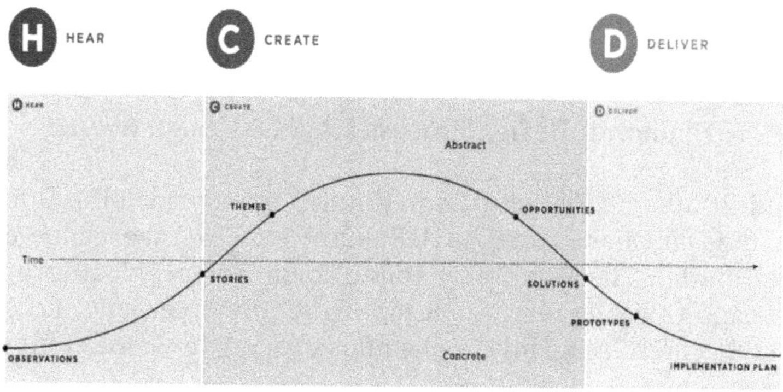

Figure 1.15 Design kit-IDEO

IDEO's 3-Stage Design Thinking Process was reinterpreted as Hear, Create, Deliver to coincide with the "HCD" acronym for Human-Centered Design.

- **Hear**: Similar to the initial stages of various design thinking methodologies, the hear stage establishes an empathic awareness of users and clarifies the issue the team is attempting to solve. It assists participants in gaining a firm understanding of the problem's background and effectively reframing it to accommodate fresh viewpoints.

- **Create**: The create stage is comparable to ideate and prototype phases in the d.school's 5-stage approach in that it focuses on exploration, experimentation, and learning through making. The problem's closest stakeholders will be further engaged in order to co-create solutions after potential research areas have been identified. As a result, design teams are able to maintain the highest levels of empathy during the early design phases and filter out any potentially harmful assumptions made by designers who do not fully get the context.

- **Deliver**: The deliver phase of the HCD process is centered on logistical implementation. It also aims to help overcome any obstacles which may exist when rolling out a solution within the required context. It is essential that solutions integrate into communities and bypass other roadblocks during implementation, and this stage will help participants achieve that.

The "Double Diamond" Design Process Model—Design Council

In the mid-2000s the British Design Council popularized the Double Diamond diagram, based on Béla H. Bánáthy's 1996 "divergence-convergence" model. The Double Diamond diagram graphically represents a design thinking process. It highlights the divergent and convergent styles of thinking involved, and is broken down into four distinct phases:

- **Discover**: The start of the project is based around an initial idea or inspiration, often gained from the identification of user needs.
- **Define**: These user needs are interpreted and aligned with business objectives.
- **Develop**: Design-led solutions are developed, iterated and tested.
- **Deliver**: The end product or service is finalized and launched into the market.

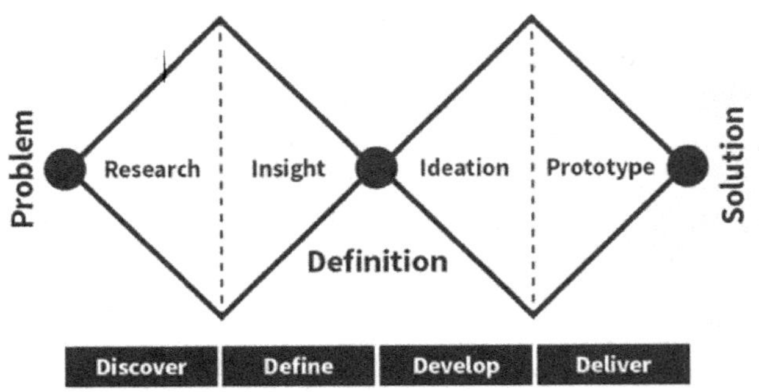

Figure 1.16 Double diamond process

The Double Diamond diagram from the Design Council helps to visualize the divergent and convergent stages of the design thinking process and highlights the different modes of thinking that designers use.

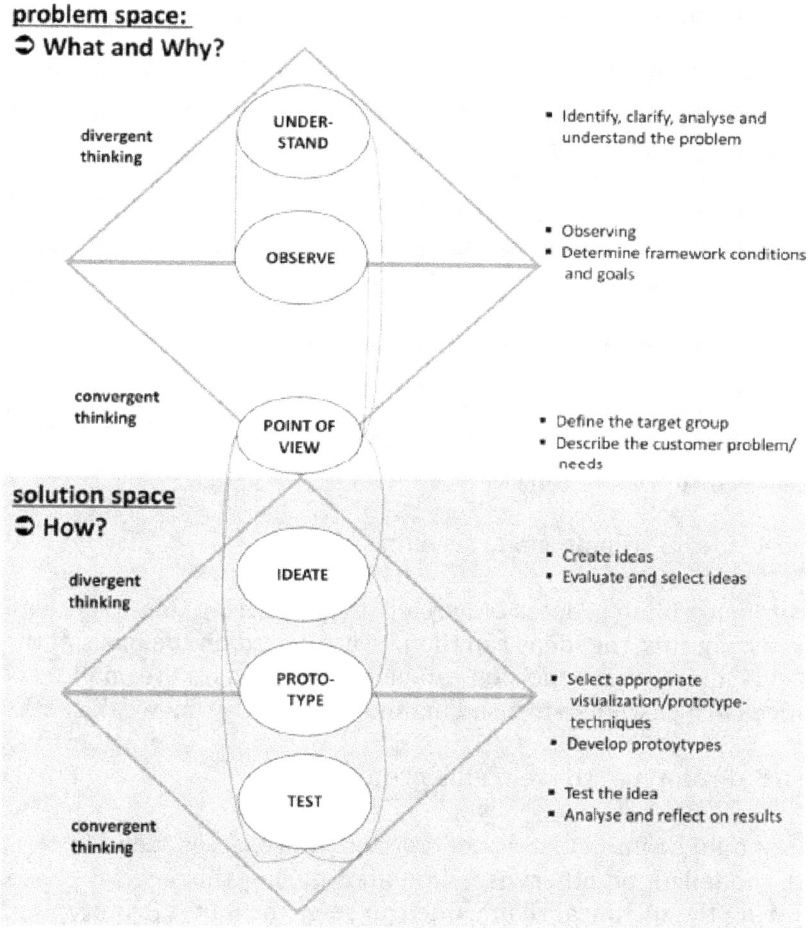

Figure 1.17 Process of design thinking supplemented with the Double-Diamond model

Phase 1 "Understand" (Understanding the Problem):

First, in the first phase, you need to comprehend the difficulty, difficulty, necessity, or requirement (problem understanding). It must be made clear who needs to be included in the process, and in particular, which technical perspective (process organisation) is required. To clarify the client's demand or problem in detail, it must be made clear how the query should be phrased.

Phase 2 “Observe“:

This stage involves thorough investigation into the requirement or issue raised by the client as well as on-site observations. This can be done using a variety of techniques, including interviews, written questionnaires, and observations that are documented with images or even movies. As a result, the general circumstances have been clarified, the target group has been precisely defined, and a thorough grasp of the customer's demands and behaviour has been achieved.

Phase 3 “Point-of-View“ (Define the problem):

Following the observations, the results should be distilled down to a single archetypal user, whose demand or problem should be encapsulated in a specific inquiry.

Phase 4 “Ideate“ (Finding and selecting ideas):

The brainstorming process only really begins during this stage. Separately from this, the ideas can then be evaluated on the basis of the idea evaluation, and a decision can be taken based on the analysis of the ideas in a customer-focused manner to identify any weak spots.

Phase 5 „Prototype“ (Develop the prototype):

Ideas should be immediately concretized, represented, devel
oped, modelled, or otherwise simulated during this crucial phase. Following the technical realm, one can refer to "Rapid Prototyping" in this context, where the production of prototypes clearly relates to both products and services. There are numerous techniques for developing prototypes that can be used for this.

Phase 6 “Test“:

The ideas will be refined and evaluated in this last stage through additional research and user feedback. Important issues with development, production, and the market must also be clarified. The real implementation phase, which involves turning an idea into a commercial good or service, would come later in the process flow described here.

Design thinking, in its entirety, is a very thorough, user-centered strategy that methodically employs methods for observation, questioning, and brainstorming as well as other moderation approaches in the many phases of a process with numerous iteration loops.

1.14 Principles of Design Thinking

When carrying out the Design Thinking process described below, the following principles must be observed, quasi the "Ten Commandments of Design Thinking":

1. Leave titles at the door!
   There is no hierarchy during a Design Thinking workshop.
2. Encourage wild ideas!
   Let your imagination run wild. Any (supposedly) crazy idea and every idea should be treated equally.
3. Go for quantity!
   Quantity before quality. Selected, analyzed and evaluated later.
4. Build on Ideas of others!
   There is no copyright. Ideas from others should be taken up, supplemented or changed.
5. Think human centered!
   Design Thinking is first and foremost thinking about people and not about technology or business goals.
6. Be visual and make it tangible!
   Use drawings, illustrations, photos, videos, prototypes, etc.
7. Avoid criticism!
   Idea generation and evaluation must be strictly separated.
8. Fail early and often!
   Failure means learning. Often failure means that you have learned a lot.
9. Stay focused!
   Set yourself limits, stick to the concrete tasks in the Design Thinking process
10. Let's have fun!

Developing new ideas in a team should be fun. Creativity needs this fun.

Summary of design thinking process and main challenges:

<table>
<tr><th colspan="2">Design thinking process</th><th>Main challenges/limits</th></tr>
<tr><td colspan="2">Inspiration–exploration</td><td rowspan="6">- Users have difficulty anticipating their needs;
- Incremental improvement of what already exists;
- Preserves the political, economic, and physical status quo;
- Iterative and explorative approach of design thinking clashes with "the linear, efficiency-oriented processes of incremental product development;
- Failing and learning from mistakes were perceived as challenging for companies with a strong risk-averse culture;
- Cultural elements of design thinking, such as having fun or interacting with users could be challenging in organisations.</td></tr>
<tr><td>Empathize

Define</td><td>- Empathy towards users to gain insight on their needs through ethnographic research;
- Challenge the initial problem/ expand problem formulation/ avoid jumping to initial solutions.</td></tr>
<tr><td colspan="2">Ideation</td></tr>
<tr><td>Ideate</td><td>Process of generating and developing ideas.</td></tr>
<tr><td colspan="2">Implementation–conceptualization</td></tr>
<tr><td>Prototype/Test/ Iterating</td><td>Rudimentary prototypes (drawing, cardboard, digital mock-up, 3D printing).</td></tr>
</table>

Module 2

Defining the Problem

2.1 Introduction

You summarize your user-related observations from the empathize stage in the define stage. Your team's work will be guided by a clear explanation of your problem statement, which will also jump-start the ideation process (third stage). The five steps don't usually happen in order; they can frequently happen simultaneously and be repeated iteratively. They also don't always have to be sequential. As a result, rather than being seen as a series of processes, the stages should be viewed as various ways that contribute to a project.

2.2 Need-finding

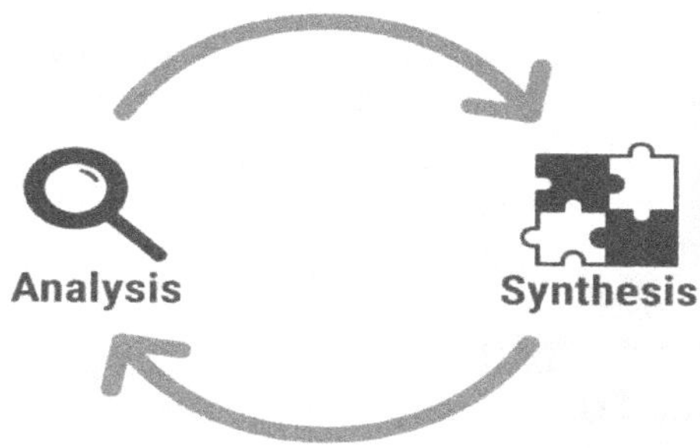

Figure 2.1 Need finding

It's helpful to first grasp the relationship between analysis and synthesis that many design thinkers will go through in their projects before we get into what constitutes a strong problem statement. In his book Change by Design: How Design Thinking Transforms Organizations and Inspires Innovation, Tim Brown, CEO of the global design consultancy IDEO, asserts that analysis and synthesis are "equally important, and each plays an essential role in the process of creating options and making choices".

The goal of analysis is to reduce complicated ideas and issues to their simpler, more understandable parts. For instance, we do it when we observe and record specifics that pertain to our users during the empathize stage of the first step of the Design Thinking process.

Conversely, synthesis entails imaginatively putting the pieces of the jigsaw together to create complete concepts. This occurs when we organize, interpret, and make meaning of the material we have gathered to formulate a problem statement during the Define step.

Although analysis occurs during the empathize stage and synthesis occurs during the Define stage of the design thinking process, these two processes are not limited to those stages. In fact, across all phases of the Design Thinking process, analysis and synthesis frequently occur in succession. Before synthesizing new insights, design thinkers frequently analyze a scenario. They then re-examine their synthesized results to produce more in-depth syntheses.

The following problem analysis, according to Kepner/Tregoe, compares the problem with a condition in which the problem (surprise) does not occur. It is based on this list of questions. The distinctions between the problem and the comparison case must be systematically examined. On the basis of this, hypotheses that explain the origin of the issue or the absence of the issue in the comparison case are developed and put to the test.

Initial perceptions of potential remedies can be gathered in addition to the problem's description. On the one hand, this entails defining the framework's parameters: What is allowed? What is possible? What is offered? Also think about what would happen if the framework's parameters changed, such as the resources that are available or required, the technical possibilities, or the political or legal environment.

On the other hand, you should carefully analyze the current state of the art:

- Why are the existing solutions not sufficient?
- Where are their limitations or shortcomings?
- Why are there no adequate solutions so far?

Table 2.1 Questionnaire to clarify the problem

	problem	non-problem	solution
Who	has the problem? is indirectly affected? believes that they are affected? makes decisions?	is not affected by the problem?	could use the solution as well? can contribute to solving the problem? does not want the solution? could stand in the way of the solution?
Where	does the problem occur?	does not the problem occur?	has something similar already been successfully resolved? is the best place to solve it? could the solution also be used?
When	did the problem start? does the problem occur? does it become an even bigger problem?	does not the problem occur?	should the solution be available? will it improve?
What	Is the problem? do you know or don't know about the problem? is not understood about the problem? is different than it should be? is particularly noticeable? annoys you about the problem? are the individual aspects of the problem?	is not the problem?	has been made the solution so far? should the solution necessarily be able to do? are the constants that must not/cannot be changed? is needed for the solution? will be different in the future? is (or is not) important for the solution? are your goals for the solution? do you have to discover?
How	does the problem manifest itself? is it related to another problem? can it be formulated differently?	is it usually going?	should the solution look like? is it tried to be solved so far? could the problem be an opportunity?
Why	is it a problem? is the problem unusual?	isn't it a problem for others?	is the solution needed? do we want to solve it? won't it just solve itself? can it be solved? is it difficult to solve?

After clarifying the problem, it is helpful to reflect again on what we know or don't know about the problem. According to Gray et al. (2010), the following matrix with the so-called blind spot of knowledge is helpful for this. The next questions have to be answered in the individual fields:

- What do we know about the problem? Which means we're aware that we know it. (known Knowns)

- What do we know that we don't know about the problem? Which means we're aware that we don`t know it. (Known Unknowns)
- What do we know without even knowing that this knowledge could help us with the problem? Which means we're not aware that we know it. (unknown Known)
- What do we know that we don't know we don't actually know? Which means we're not even aware that we don't know. (Unknown Unknowns).

2.3 Analyzing Product Opportunity Areas

A method for determining how desirable a business opportunity is market opportunity analysis. Frequently, customers approach us with what they believe to be a superb marketing or product concept. While it's a terrific idea, research reveals that they came to the market too late and so lost the window of opportunity, even though it is a brilliant idea. There are 6 key questions we get answers to when doing market opportunity analysis:

1. What type of an opportunity is this?

    - An unmet customer need
    - An improvement over a product/service already on the market
    - Disruptive idea or technology
    - Horizontal improvement which will improve functional operations
    - An opportunity that is targeted to a specific industry segment

2. How are people meeting this need today?

    - Direct competitors
    - Indirect competitors
    - Substitutes
    - Collaboration
    - Think Porter's 5 Forces

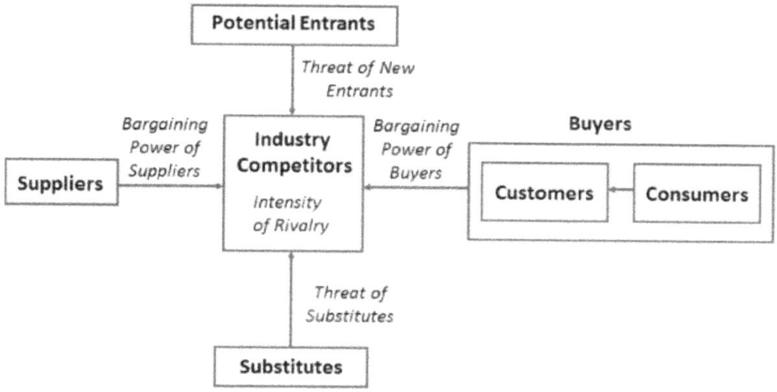

Figure 2.2 Market opportunity analysis

3. Is another company already filling this need?

4. If so, how can I differentiate my company's product or service to create a competitive advantage? If not, is there sufficient market demand?

5. Is this opportunity financially viable both to our customers and to our company?

- How will this impact cash flow?
- What investment do we need to make to be in this space?
- What about distribution channels?
- Will we need to train our sales channels to market this opportunity?
- Will we need to establish partner relationships to market this product or service?
- Can we market this to our existing customer base or will we need to chase different customer segments?

6. What external factors should we be considering that might impact this opportunity?

- Think STEEP analysis: social, technological, economic, ecological, political/legal

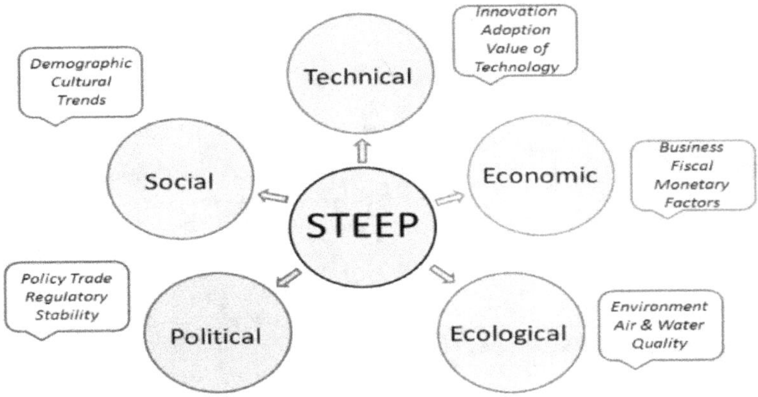

Figure 2.3 STEEP analysis

2.4 Market Analysis

A market analysis evaluates a market both quantitatively and qualitatively. It examines the market's size in terms of volume and value, the different consumer categories and purchasing trends, the level of competition, and the overall economic environment in terms of entry obstacles, laws, and regulations.

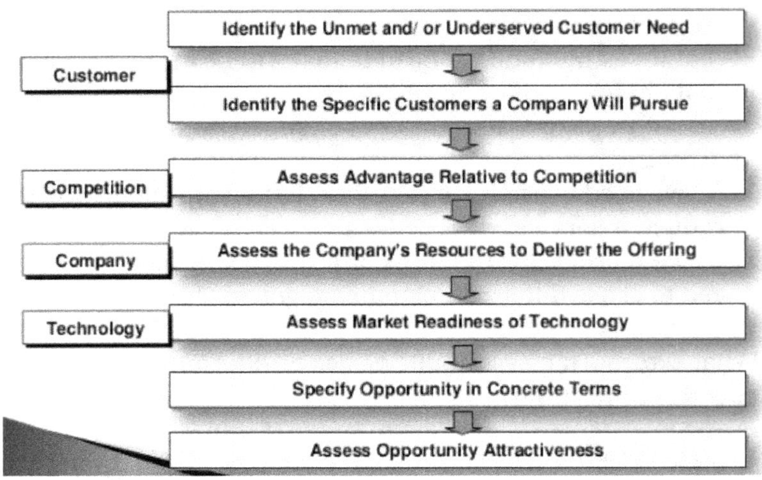

Figure 2.4 Market analysis

The first step in figuring out if there is a demand or audience for your concept, whether you are starting a new company or launching a new product, is to do a marketing study. You may build your product or service and marketing strategy using vital information that you gain from understanding the demands of the market and how it is currently catered to. A "novel" concept with a small market is all too frequently launched by enterprises at a cost of thousands of dollars due to competition. The owner is compelled to rethink his plan of action and decide whether there is place for another player.

Although the product's quality is essential, having the finest product on the market does not guarantee that you will have the most sales. A product's pricing might include up to 50% of marketing costs. The business that succeeds in the marketing game will often earn a higher market share. See Identify Your Target Market for further information.

Conducting a market analysis will help you:

1. Prepare to enter a new market
2. Launch a new product/service
3. Start a new business

Although "marketing" and "marketing analysis" are both information games, they should not be used interchangeably. All the activities involved in promoting a good or service are included in marketing. A marketing analysis is the real evaluation of the target market, competitors, and marketing requirements for that good or service.

The marketing analysis process can be broken down into six steps:

1. Defining the problem
2. Analysis of the situation
3. Obtaining data that is specific to the problem
4. Analysis and interpreting the data
5. Fostering ideas and problem solving
6. Designing a plan

1. Defining the Problem

Defining the problem is crucial to conducting a successful marketing analysis. This may require a great deal of time but it is well worth the time and energy expended. Defining the objectives is tantamount to a successful marketing campaign. Many individuals waste valuable time performing good research on the wrong problem.

The following questions will assist you in defining the problem:

1. Are we trying to market our entire product or service line? Or, are we trying to hone in on a new product or a new service?
2. What specific marketing strategies have we utilized in the past two years?
   a. How has each strategy affected sales?
   b. What strategies are we currently using?
   c. How do our competitors market their product?
3. How much money is allocated to marketing?
4. When making a sale, do we survey our customers to determine a referral source?
   a. Do we thank our referral source?
5. Why would someone choose our product?
   a. What differentiates our product from our competitors' products?
   b. Why do people choose our competitors' products or services?
   c. Do we need to enhance our current product or service?
6. Who are our customers?
   a. Are they from a specific region?
   b. How do we attract new customers?
   c. How do we increase sales from current customers?

Many businesses attempt to address sales-related issues without first determining the nature of the issue. A famous instance happened in the soft drink business when Coke and Pepsi, the two leading competitors, stepped up their marketing efforts to compete for greater sales. At the same time, Dr. Pepper's sales started to fall, which was blamed on a subpar advertising effort. The "Be a Pepper" phrase was then implemented. Despite the success of the advertisements, sales kept declining.

The marketing managers were forced to revisit their issue. Focused study showed that Dr Pepper's target market was distinct from Coke's and Pepsi's. According to research, Dr. Pepper's target market feels that life should be lived in line with one's own set of personal values rather than in accordance with others' expectations. Conversely, those that consume cola do so to gain favor with others. Despite the memorable phrase, Dr. Pepper's marketing team drove away potential customers with a campaign that asked them to join a sizable community of "Peppers."

The soft drink industry clearly exemplifies the importance of defining the problem. It is important not to confuse the symptom with the problem. If a company is having trouble selling a certain product, it doesn't mean that there isn't a need. The problem may be distribution and/or pricing. Mistaking symptoms for problems will lead to misguided research and serious mistakes.

2. Analysis of the Situation

An analysis of the situation is an informal survey of what information is available in the problem area. The analysis will help define the problem and ascertain the need for additional information. This process entails informal talks with informed people. Informed individuals can be others in the company or outsiders with knowledge about the industry or product. In some instances, customers are contacted to provide information.

The analysis process is crucial in situations when the marketing manager lacks context. It's critical to comprehend the problem area, including the target market's makeup, the competition, the marketing mix, and the surrounding environment. Without this understanding, mistakes might be expensive. This issue can arise for a shop that wishes to poll his consumers. In-store interviews will be conducted by a research company. For instance, the hired company is unaware that several of the establishments are undergoing renovations. As a consequence, the data gathered shows how much attention customers pay to aesthetics, noise level, and trouble finding things due to construction. The knowledge would be useless.

The analysis should focus on both primary and secondary research.

Primary and Secondary Market Research

If you don't already know the answers to all the questions mentioned in the Problem Definition section, you can perform primary research or access secondary research to get the information you need.

Primary research is analysis that is deliberately produced with a particular goal in mind. Focus groups, qualitative surveys, and phone interviews are all examples of primary research. You gathered this knowledge on your own.

Secondary research, on the other hand, refers to study that has previously been done for another reason. It contains important information that may be obtained. You may find secondary research at libraries, online, in journals, books, etc. The Internet is the quickest and most effective way to access this kind of information.

Library information can also be found on the Internet. Access your favorite search engine and type in the following:

- For businesses Type in the trade association name followed by either ".com" or ".org"

- For universities Type in the name of the institution followed by ".edu"

- For government Type in the name of the department followed by ".gov"

Simply enter "industry research reports" into a search engine to get a range of websites if you want to perform a thorough industry search. Your search will be more focused the more exact you are. An important resource for doing a competitive analysis is the Internet. Just enter the business name and the dot com after it. If that doesn't work, try going to the internet yellow pages and typing in the name of your rival's firm. You can get an inside look at their services and several other reports if they have a website.

Demographic and competitive reports can be attained for a fee at Dun and Bradstreet's Web site, dnb.com.

Industry newsletters can be emailed to you when new data is reported. By searching your industry on the Internet, you will uncover lots of potentially valuable marketing analysis information.

Other widely used standard research sources include:

- Trade and Industry Sources such as Gale Research publishes the Small Business Source Book, which has a strong focus on retail trade. It lists industry associations, trade shows and conventions, consultants and venture capital firms.

- Forrester Research and Gartner Research publish detailed reports and studies focusing on the information technology industry.

Make sure your market will accept your goods or service before you enter it once you feel confident doing so. If your target market doesn't think your widget is the most inventive on the market, you might be in for a pricey year.

The best kind of primary market research study is thorough. The study should incorporate findings from laboratory and field investigations as well as focus groups that were conducted by experts if it is to be complete. Conduct a focus group to get a sense of comfort if you can't carry out a thorough study.

3. Obtaining Data Specific to the Problem

The following phase is gathering primary data and carrying out a proper research effort. Primary data collection can be done in a variety of ways. The goal of the research is to determine what consumers believe about a subject or set of behaviors. You can conduct research in person or online. Quantitative or qualitative questions are both acceptable. Observing consumers as they make purchases or utilize a product or service is another research strategy. Open-ended inquiries are used in qualitative research to elicit detailed responses. Avoid asking closed-ended questions with yes-or-no options. Without providing them with detailed instructions or rules, the goal is to get individuals to voice their opinions on a subject. These are some illustrations of qualitative research questions:

"What do you think about when you decide on a place to shop for your kids' clothes?"

The consumers are allowed to respond anyway they like; some may discuss the handy location, others the service, and yet others the types of designer items that are offered. The key component in this kind of study is depth. Each person might be questioned further to learn more about their reaction and purchasing behavior. When summarizing all the data, the qualitative technique calls for the researcher to use discretion. The key is depth.

Question-oriented qualitative research is not required. The most popular method for conducting qualitative marketing research is focus groups. In an informal atmosphere, six to ten persons will be interviewed for this. To learn more about the topic in-depth, open-ended questions are offered. In a group setting the researcher looks for group interaction to stimulate thinking.

This strategy may teach a focus group leader or facilitator a lot. The facilitator's duties include establishing ground rules for group engagement, speaking as little as possible, maintaining group concentration, and asking straightforward, open-ended questions. An hour-long focus group session is common. The meetings can be recorded on video for viewing by several management. The conclusions drawn differ according on the person's level of expertise. Qualitative research can be impartial, but it takes substantial education and expertise to do.

Qualitative research may provide ideas or hypotheses, but other approaches based on larger sample size and objective measures are needed to test the hypothesis. Researchers often use qualitative research to prepare for quantitative research.

In contrast to qualitative research, quantitative research collects parametric statistical data, or data with a number attached. In general, sample sizes are bigger and more accurate representations of the market.

There are inferences that may be derived from the statistics or data collected. Typically, survey research is quantitative in character. It looks for organised replies that may be reduced to a few figures, such as averages, percentages, or other useful data. An illustration of quantitative research is the proportion of shoppers who buy coffee at grocery shops. It is possible to compute an average score.

For the sake of simplification, survey questionnaires frequently include predetermined answers to questions. The respondent can complete it more quickly and easily because to the multiple-choice style. There are also many more questions with simple fill-in-the-blank, circle-the-range, or precise answer options. A market researcher can inquire about your wage range or the number of vendors you use for home appliances (multiple choices given with ranges of salaries). Most surveys are evaluated using computer-friendly fixed replies.

The attention of a respondent may be successfully maintained in surveys through personal interviews. In this style, the interviewer is there to clarify challenging instructions in the hopes of getting better replies. In the retail and industrial settings, this style of interview is typical. Interviews are occasionally conducted on-site to save money on the expense of finding customers. A random selection of consumers is questioned. Having a well-trained interviewer helps to decrease the chance of biasing the response.

The survey of personal interviews has a few drawbacks. It is more expensive than a mailing or phone contact, and occasionally replies won't provide honest responses to all questions, particularly those that touch on delicate family matters.

Data gathering may be done through observation. Researchers aim to understand what a topic performs naturally via observation. Consumer behavior shouldn't be affected by observations. Research on advertising frequently uses observational techniques.

4. Data Analysis and Interpretation

Analyzing and interpreting data is essential for market analysis.

What does this information mean? Is it possible to outline the issue and then create a strategy using the data? This phase in quantitative research most frequently uses statistics. There is several computer-based statistical software available on the market for data analysis.

Samples are required since it is difficult to collect data on every member of a chosen population. The relevant population includes a sample population. Its validity depends on how accurately the sample represents the relevant population. Results from an unrepresentative sample will have an adverse effect on your marketing.

Example of Poor Sampling: During business hours, a store manager has his workers make phone calls to customers. Between 11:00 am and 2:00 pm, the survey is taken. Because those who work during those hours are not included in the poll, it does not accurately reflect the broader population. Only non-working or chance homebodies will be evaluated in the sample. Another factor to consider is the caliber of the study data. In other words, you must be measuring what you intend to measure for the results to be legitimate. A poorly phrased query may be unclear and provide false information.

Marketing managers must make sure the data supports the conclusions made in addition to sampling and validity difficulties. The step of interpretation is this. Even when the right statistical method and precise computations are used, the interpretation might be off. Example of Misinterpretation of Data: Parents were asked to rate the use of five newborn car seats in a survey. From "most favoured" to "least preferred," they were to rate the seats. Fewer people gave one car seat the top spot than any other car seat. According to the studies, parents of young children preferred it above other car seats. They neglected to mention that this particular car seat was the least preferred by 70% of respondents. So, the first-ranked car seat was most preferred by only 30 percent of the parents.

5. Fostering Ideas and Problem Solving

In this stage, marketing decisions are made using the study findings. Planning for marketing should take into account the findings. The firm, no matter how big or little, has spent time, money, and people on needless data if the study doesn't give the knowledge required to make these judgments. The final step must be anticipated throughout the entire process.

6. Marketing Plan

This six-step market study procedure is essential for creating a marketing strategy that is customised to your unique product or service. The procedure may reveal a serious but unidentified issue that was previously present. The researcher and business owner can swiftly arrive at a workable solution by identifying and concentrating on the true problem.

A marketing strategy outlines the specifics of how you'll promote or sell your item or service. To recapitulate the point of this lecture, the marketing plan's function is to provide you direction as you study your market. There are several tools and software programmes available to help you improve your marketing efforts.

Software packages are numerous and include:

- Business Plan Pro, Palo Alto Software
- Small Business Advantage, Encore Software

2.4.1 Different Methods of Market Analysis

You will need accurate information to do a market study. Small businesses often conduct the essential research for their market study alone. On the other hand, larger businesses frequently hire market research organisations to conduct it for them. Various data collecting techniques may be used to conduct a market study. Primary and secondary research are distinguished from one another.

Primary research involves interviewing subject matter experts from a target market to gather fresh data. This has the benefit of keeping your own study objectives in sight. You may get the information you want for your market study in this manner. Contrarily, secondary research makes use of data archives from earlier surveys. Both internally and externally, this can be gathered. You may save time and money by using secondary research because you won't need to perform expensive interviews and evaluations. The Federal Statistical

System, professional associations, annual reports of other organizations, or trade publications are a few examples of sources for representative data.

2.5 Product Research

Product research is the process of determining whether your idea for a new product or service might be successful and how best to develop and sell that product. This method starts with a market investigation to identify if comparable items are offered. You can decide if there is a market for your concept after you are aware of what is already offered to customers.

2.5.1 Methods of Product Research

There are several types of product research, including:

Concept testing: Concept testing involves polling consumers about your proposed product to get their opinions. Concept testing is a flexible tool that may be used to gauge consumer reactions to a variety of product features.

Price testing: The cost of a product is among its most significant characteristics. To determine demand elasticity—the relationship between a product's demand and price—you poll groups with several pricing alternatives. This process is known as price testing. Demand elasticity is a crucial piece of knowledge to assist you choose the price point that will yield the greatest profit.

Product tests: You may test-market your product in a small region before you distribute it to see how it could do. Determine which strategies get the most reaction by experimenting with alternative marketing messages, pricing points, and distribution strategies.

Focus groups: Organize a focus group with customers of the goods you sell. You can enquire from them about their opinions about the pricing, usability, and other features of your goods. To reduce the possibility of bias in your interviews, you might decide to employ surveys or conduct third-party interviews.

Product demos: Create a test model of your idea after researching the market, similar items, and possible clients. To acquire their opinions, you can show that model to a group of prospective clients. Find beta testers for your digital product who could become customers.

Product surveys: An efficient technique to find out what potential buyers think of your product is to conduct a poll. Your survey can ask respondents about their opinions on various aspects of your product as well as how their design stacks up against others on the market. These questionnaires can be completed in person, over the phone, online, or by mail.

Importance of product research: Your product has the highest chance of success through product research. You may obtain a competitive edge in the market by learning what customers want and changing your product accordingly. You may create products with a better value and introduce innovation to the market with the aid of product research. You may achieve or even surpass your company objectives with superior products.

2.6 Product Design Process

The common list of processes in the product design process and have listed them below in to consistently apply all five of the aforementioned phases. The following are the steps in the product design process:

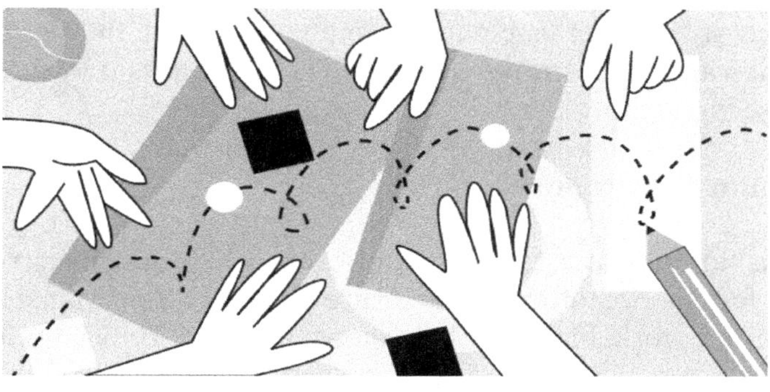

Figure2.5 Design process

1. Brainstorming

Brainstorming, which originally occurred in the United States in 1953, is the initial stage of product design. Two groups are often formed to carry it out. People who come up with solutions to the problem are included in the first group. A "commission" that handles the processing of the suggested ideas makes up the second group.

Brainstorming consists of three phases:

Definition of the problem. In particular, you must announce the problem (in this case, the need to create the "right" product design) and select participants for both teams.

Idea generation. The most inventive strategy is used during this phase. Participants in the brainstorming process express any concept they can think of no matter how absurd it may seem. Here, quantity above quality is key. Therefore, it is necessary to be guided by the following rules:

- Generate as many ideas as possible, without any limitations.
- Even the most absurd and non-standard ideas should be considered.
- Ideas can and should be combined and improved.
- No criticism is allowed - at this stage, ideas should not be evaluated in any way.

Selection of the most viable ideas among proposed. In this step, appreciation and criticism are welcome. This stage is usually conducted by the second team

2. Defining the Product

You must identify a certain number of general expectations (requirements) for the execution of the future product after brainstorming, once all the workable ideas have been selected. Typically, the outcome is a pretty abstract list, the purpose of which is to establish the direction of future team activities rather than to provide a precise guideline on the development.

3. Conducting the User Research

Research encompasses a variety of ideas, including market research to determine the presence of rivals, trend definition, the estimation of the product's potential lifespan, etc. Typically, business analysts and marketers do this work. The findings of this study give reason to create so-called customer portraits. Your team would therefore have a better concept of what the finished result ought to be thanks to these photographs. Gender, age, marital status, income level, location of residence (geography), occupation, work position, normal issues, demands, concerns, and desires, among other factors, are all factors that should be considered while composing a picture.

You must never skip this stage in an effort to provide a product "for everyone," regardless of cost. With a high probability, you simply will not please anyone and drain your entire budget "into the pipe".

4. Sketching

Sketches should always be produced for any huge project with a sizable budget. The search strategy must be agreed upon with the customer before any time is spent looking for answers. When picking the main concept, sketches enable you to slightly limit down this particular route. Work with composition, layouts, edits, concept refining, etc. would then come next.

5. Prototyping

Making a prototype is just as necessary and important as designing the design itself. Be aware that the outcome is only a "skeleton" of what the final product will look like rather than a stunning final appearance. Nevertheless, it enables showcasing the product's functionality, modes of user engagement, and fundamental design. By building a prototype, many future mistakes and fixes can be avoided. Both the consumer and the performers will benefit from the time, money, and nerve savings.

6. Compiling Specifications

The process of elaborating the aforementioned need specification really entails the creation of a list of specifications. It provides for the most thorough examination of all demands for the end product and potential remedies. These specifications should also include the precise division of tasks, due dates, and expenses. At the time of product development, these records would be crucial.

7. Producing the Factory Samples

You may determine whether the main concept of the product would be practical and appealing to actual users by producing pre-production prototypes, or whether it could be wiser to change the activity vector to other important areas. It also enables knowing how much each production department or contractor is prepared to create the designed items at the needed rate and level of quality in the event of a physical product.

Such samples are known as MVPs in the context of software; they are a type of wireframe program that only includes the most fundamental features.

8. Sample Testing

The ability to test samples enables you to quickly spot product faults or compliance issues. That is before you even start the generally costly task of putting it into practice. As many times as necessary can be spent producing and testing the samples to ensure that they meet all specifications and receive enough favorable consumer feedback.

9. Starting the Production/Development

You will already have a complete set of technical specifications with precisely specified needs, tasks, responsibilities, dates, and budget when you begin manufacturing or creating the product. Your objective is to assign priority and divide the large activities into smaller subtasks (so that the timeframe for completing them does not exceed several weeks). Agile approaches like Scrum or Kanban are used for this in the software business nowadays.

10. Providing Quality Assurance

Quality assurance efforts include the release and subsequent maintenance phases of product development. Their primary duties, however, were to guarantee the caliber of the final output. Pre-release testing is carried out in the case of app development to check the quality of the delivered solution, its compliance with the requirements specification, and the target audience's expectations (they are determined at the stage of forming the customer portraits).

2.6.1 How to Evaluate Product Design

We advise performing a SWOT analysis to get a neutral evaluation of the final product design. It suggests solutions to queries in four areas, in particular:

Figure 2.6 SWOT analysis

What benefits does your product offer? How readily this value can be communicated to customers, which user issues it resolves, how closely the deadlines were adhered to, etc. are just a few examples of how beneficial the implemented functionalities are for the target audience.

What drawbacks does your product have? (Are there any ways to make your concept better, what resources does your team need for its execution, what challenges could arise in successfully promoting it to the intended audience, what drawbacks exist from the perspective of the intended audience, etc.)

What opportunities exist for promoting your product? (For example, how straightforward it would be to promote it on the market, what are its competitive advantages, which channels of marketing would be most effective, how successful it will be over the long run, etc.).

What dangers can there be to your product's effective promotion? (Would the product's drawbacks pose significant risks for marketing it; what potential financial issues would you face; are there any rival items that are difficult to surpass; etc.).

Changing/Modifying Your Product Design

You must adhere to the aforementioned product design phases in order to produce a product design. But if you want to change the final design, you'll need to add two crucial steps first.

Understanding your goals and what the modified design may assist you accomplish is the first step. For instance, if we are discussing a website, you need to look at the measures that determine how popular it is: traffic, conversion, user retention, etc. As a result, before determining if changes are required, you must first study the indications of the site's current version. A/B testing can also be used to determine how much alteration is necessary. You may make some adjustments to the current design, observe how users react, and determine whether it would be just as beneficial as creating a new one.

The following step is to conduct an end-to-end competitive analysis. It is critical to evaluate the expenses of the competing products as well as their real qualities (it may be so that the budget of altering the product design to overcome some competitors would significantly exceed the expected earnings). Then, you may apply those elements that have shown to be appealing to target markets like your own.

The design phase is the following step in the product strategy. Product teams should adhere to a methodical procedure known as the design process while creating new products.

The following are these steps:

1. Define the Product Vision

A product vision, or product vision statement, describes the overarching long-term mission of your product. Vision statements are aspirational and communicate concisely where the product hopes to go and what it hopes to achieve in the long term.

Your product's overall long-term objective is outlined in your product vision, often known as your product vision statement. Aspirational and succinct, vision statements explain where the product aims to go and what it expects to accomplish in the long run.

A strategy outlining how to accomplish your goal shouldn't be part of your product vision. Instead, you ought to maintain the product strategy, or the means by which the goal will be reached, apart from the product vision. As a result, you can alter your strategy while still remaining true to your original intent. (Lean Startup refers to this as pivoting.)

For (target customer)
Who (statement of need or opportunity)
The (product name) **is a** (product category)
That (key benefit, reason to buy)
Unlike (primary competitive alternative)
Our product (statement of primary differentiation)

Figure 2.7 Product vision

It's not only about evoking feelings in your potential customers when it comes to your product idea. It is crucial to the regular decision-making procedure carried out by your team. It directs and reroutes everyone on the product team toward what is most important to the business.

A deliberate and well-crafted product vision, in the ideal world, may motivate your team to pursue excellence in the job they do each day. When the product vision is done properly, it may serve as a reminder to each member of your product team of how their work fits into the overall goal.

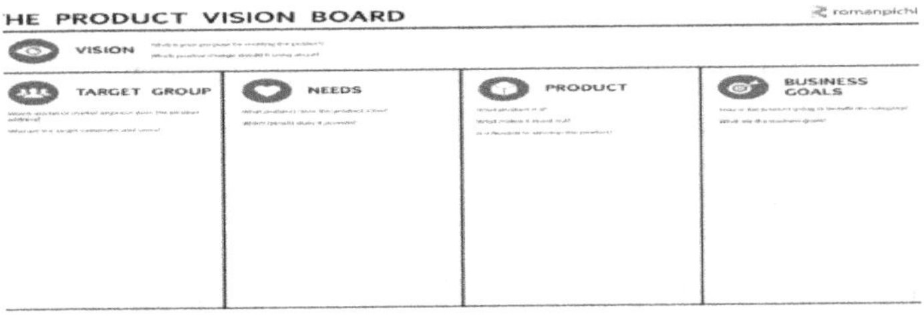

Figure 2.8 Product vision board

At the outset, try to understand the product vision and how it will help the users. Consider its purpose and be certain of the objectives it will assist in achieving. A goal without a deadline is but a wish, after all. The same is true for product design. After defining the product vision, create the product strategy by mapping out the path to the desired outcome.

The long-term results and advantages you want to see from your product should be the main emphasis of your product vision. Strategize: Product Roadmap and Strategy your "vision reflects the ultimate motivation for generating the product," according to Practices for the Digital Age.

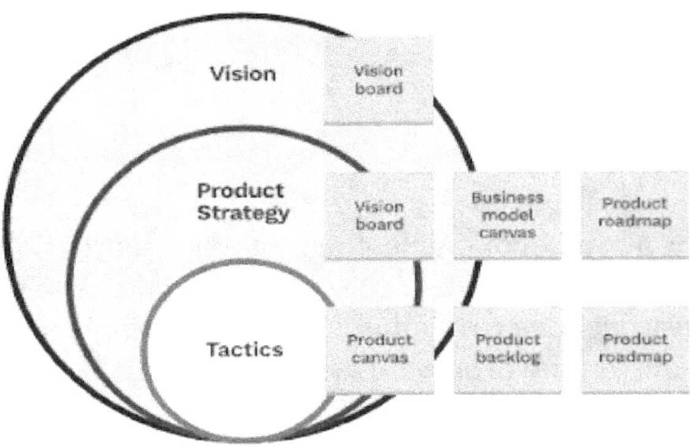

Figure 2.9 Product strategies

In general, you need to have a few basic elements for good product strategy planning. It includes:

Market and needs: Understanding the target audience and the problems they face that you wish to solve in a way your competitors haven't done before

Key features and differentiators: What your competitive differentiation is and how you'll deliver that value to your customers

Your product strategy needs to be flexible and adaptable. Understand that your team will always be learning about the users and the product as it evolves and makes its way down the lifecycle.

Something you can do to make sure you have a solid product strategy is to ask yourself, and your team, the following product strategy questions. The answers should have been the foundation for your strategy back when you first built it from scratch, but they can be used to calibrate it:

- What are the emotional reactions you're hoping to get from your users every time you release a feature?
- What unique purpose will your product fulfill in the market?
- What are the valuable aspects you need to build into your product?
- What resources (time, effort, money) does the company have to achieve that value?
- What are the limitations and possible snags your product might face in the market?

Research About the Users

User research is the next stage in the product design process. Proper research will act as a motivator in persuading the stakeholders to adopt your suggestions. Ask the appropriate questions during user interviews to gather information. This will reveal how the user sees his or her own requirements. Additionally, it will reveal your degree of attentiveness. As an alternative, you may design online polls and surveys and do a more thorough market analysis.

Analyze the Users

You should analyses the data gathered when the user research is finished. Draw inferences from the buyer personas and create an empathy map to help you understand what the user says, hears, thinks, and feels.

Your product plan outlines how you'll get there if your product vision is the long-term result you intend to achieve with your offering. Your strategy serves as a connecting point between your roadmap and your product vision, converting the latter into actions you can take to meet your long-term objectives.

Product managers can define their product strategy with the aid of the product strategy canvas, which is a step-by-step manual. You may be sure you're on the right track to establishing the factors that are most important to product success by answering the many questions and sections of a product strategy canvas.

The product strategy canvas developed by Roman Pichler is a useful tool for defining your target customer, their problems, and how to translate those problems into tasks that the team can work on.

Figure 2.10 Product vision statement

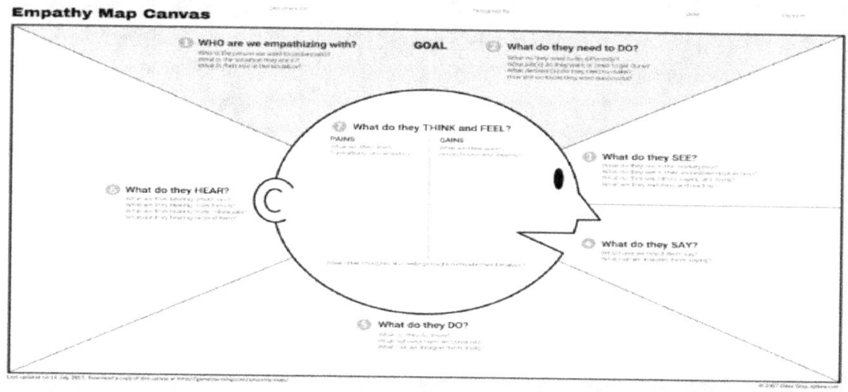

Figure 2.11 Empathy map canvas

Ideate

Together, the team members come up with a ton of original ideas that might assist realize the project's objectives. Here are some tips to make the brainstorming process easier.

User journey mapping: It includes a path that depicts the user, his or her goal, and a scenario. In case you have a weather app, your user journey map would go as depicted below:

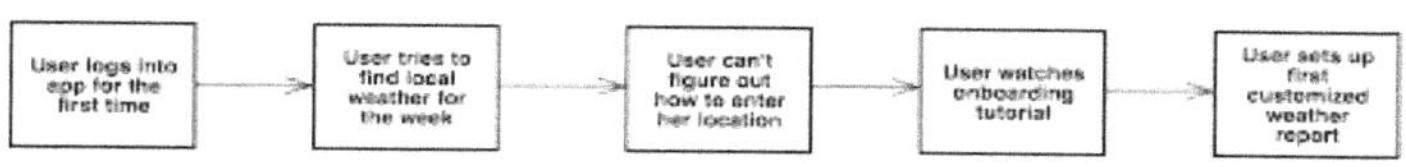

Design

The product team should be certain of their goals at the conclusion of the ideation process. The phase will start with the development of the solution to the client's issue.

We will discuss the idea of "Minimum Viable Product" before getting started on the product design.

Make sure a product's main features are as basic as possible while designing it so that it can be used. The term MVP refers to these qualities.

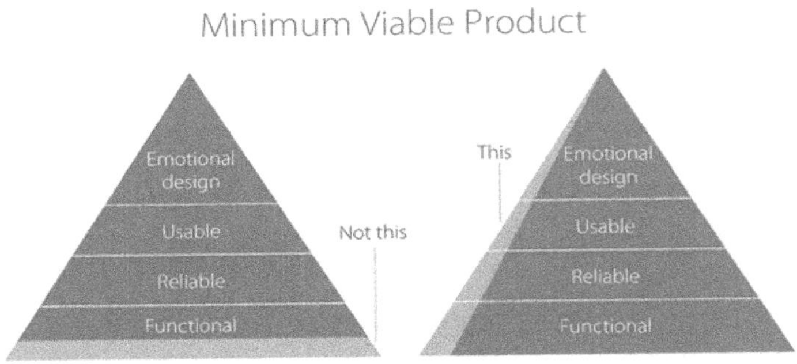

Figure 2.12 MVP

2.7 Framing a Generative Challenge Statement

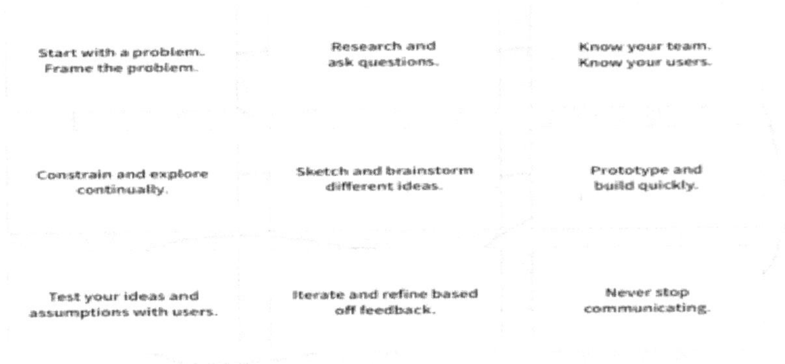

Figure 2.13 Generative challenge statement

Before examining potential solutions, it is a method used to validate and explore the issue space in order to arrive at an actionable problem statement. It is a crucial phase in the issue-solving process since even little modifications in how a problem is first characterized can have a big influence on the eventual solution.

Time spent on this process up front results in better solutions that are supplied more quickly and for less money.

The list of low-effort, high-value activities our clients may implement right now to advance toward their goals is a useful byproduct of our one-day training. Additionally, the workshop determines which initiatives are high-effort but low-value and should be set aside for the time being.

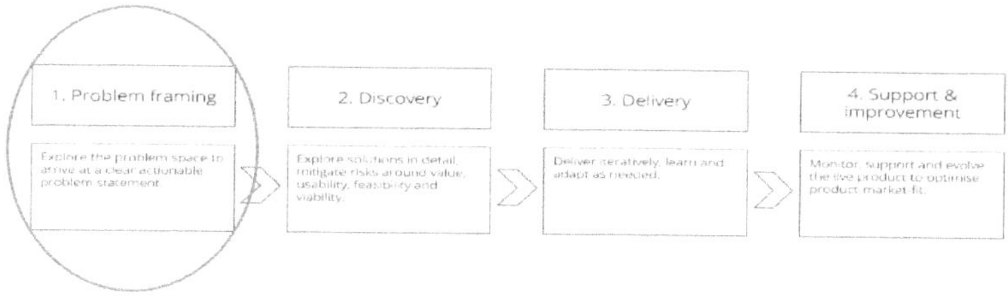

Figure 2.14 Time spent in problem framing

Step 1 - Create a Map

A basic, high-level map of the company workflow is what we start with. This is helpful in two ways:

It immediately piques everyone's interest in the room. Usually, will one individual be fully knowledgeable about each phase.
It offers a helpful perspective for understanding where in the process or company issues arise. This is crucial since problems frequently have their fundamental causes one or two steps upstream from where they are first seen.

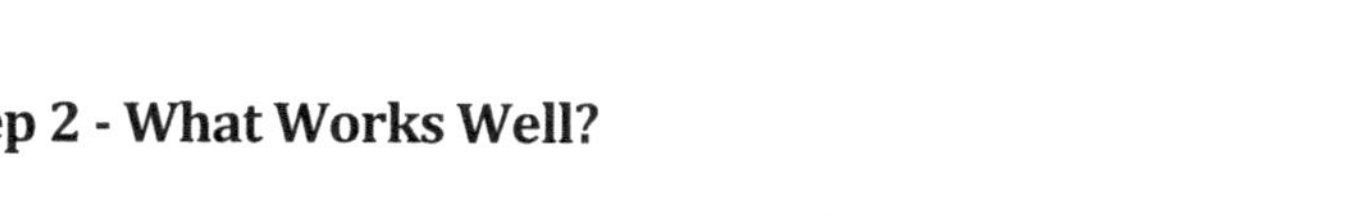

A simplified map of a workflow

Step 2 - What Works Well?

We then take some time to concentrate on what is going well. Though it might be tempting to jump right into problem-solving mode, remembering instances in which we have successfully solved problems in the past creates the conditions for original creative thinking and fosters a "can do" attitude.

Individually, each person adds post-it notes to the process with examples of things that perform effectively. The end result is a precise visual representation of the company workflow's successful regions.

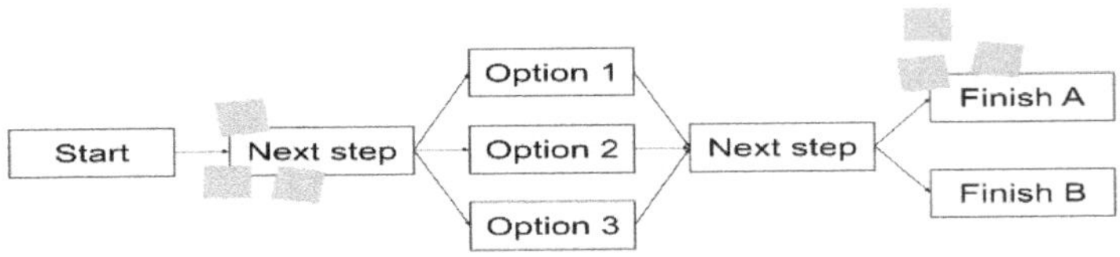

A workflow with success areas pinpointed

Step 3 - What are Our Problems?

Once we have drawn out our procedures and celebrated our triumphs, we're ready to add the difficulties. Again, working alone, each person identifies and adds any difficulties, worries, or bottlenecks to the workflow we have laid out. Problems that are similar are grouped together.

Because difficulties are simply chances for progress, I normally advise everyone to utilize green post-it notes in this situation. The result is a map of opportunities that are just waiting to be taken advantage of.

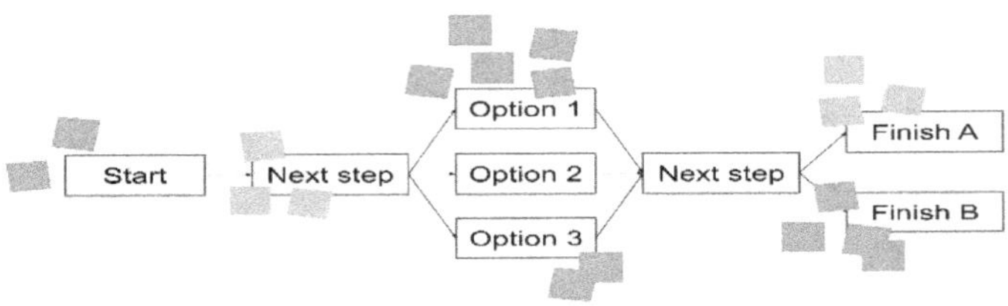

A simplified map of a workflow with yellow and green post-it notes

Step 4 - Priorities Problems

We will spend the afternoon delving deeper into the top issues. This indicates that we must organize our tasks in order of importance.

According to our experience, one afternoon is exactly enough time to thoroughly examine two or three problems and come up with a variety of viable solutions. Dot voting is a useful tool for keeping the group on course without getting distracted by pointless arguments and snide remark.

Everybody casts a vote for the issues and challenges they believe are most crucial using their dots. They can freely vote more than once and on their own proposals. Typically, the number of choices a person votes on determines how many dots they receive: more choices = more votes.

The top two or three problems advance to the following round when the votes are totaled. The remaining issues are listed for potential future attention. The senior decision-maker in the room casts the deciding vote in the event of a tie.

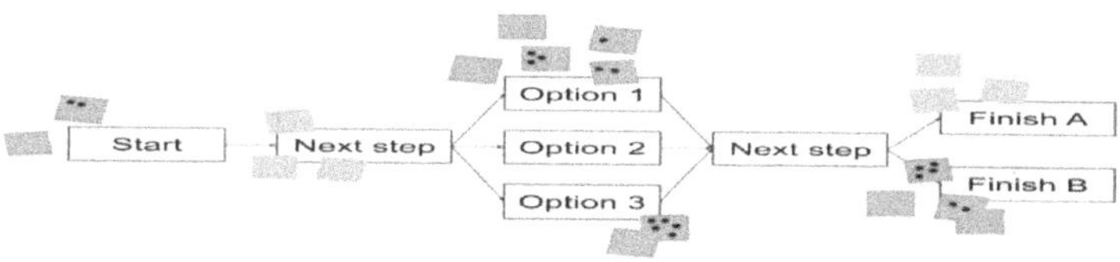

A workflow after problems have been prioritized

Step 5 - Assess the Impact

Solutions need to work for the business and the customers who buy its products and services. It is therefore vital to take some time to view the problem through both of these lenses.

Step 6 - 5 Whys

Before ideating solutions, we do a quick exercise to help us get to more in-depth insights into the problem we're focusing on. We use the Five Whys technique to investigate the root cause.

Asking 'why' lets you explore the underlying cause-and-effect relationships of a particular problem. For example, a problem statement could start as "Improve the customer support ticketing system".

Asking 'Why' may have the following results:

Why? Because support tickets take too long to resolve.
Why? Because lots of new tickets are created each day.
Why? Because all 80% of new customers submit a ticket.
Why? Because they need help setting up their account.

By doing this exercise, you may turn your attention away from a problem's surface symptoms and onto its root causes. Including "How may we" at the appropriate level positions us to develop useful, all-encompassing solutions that meet business, and user needs as well as deal with underlying problems.

Step 7- Explore and Prioritize Solutions

I frequently doodle on a piece of paper or a whiteboard. It's like speaking out loud, but with pictures.

We created our workshop keeping in mind that while human short-term memory isn't so great, spatial memory is fantastic. The space becomes the team's collective brain by plastering the walls with ideas and work throughout the day. When it comes to the day's last stage, which is developing and prioritizing solutions, this is really helpful.

2.8 HMW Question Framing

Figure 2.15 HMW question framing

You may begin to come up with solutions to your design problem once you've characterized it in a POV. Asking a particular question beginning with "How May We" or "in what ways might we" can help you begin to use your point of view. How Might We (HMW) inquiries have the ability to start brainstorming sessions and other ideation processes. They must be both broad enough to support a variety of solutions and specialized enough to support the development of tailored solutions. The observations you've obtained during the Empathize stage of the Design Thinking process should serve as the foundation for your "How Might We" queries.

For example, you have observed that youths tend not to watch TV programs on the TV at home, some questions which can guide and spark your ideation session could be:

- How might we make TV more social, so youths feel more engaged?
- How might we enable TV programs to be watched anywhere, at any time?
- How might we make watching TV at home more exciting?

The HMW questions open up to Ideation sessions where you explore ideas, which can help you solve your design challenge in an innovative way.

Why-How Laddering

"As a general rule, asking 'why' yields more abstract statements and asking 'how' yields specific statements. Often abstract statements are more meaningful but not as directly actionable, and the opposite is true of more specific statements." - d.school, Method Card, Why-How Laddering

Designers attempt to describe the problem at the Define stage and typically inquire as to why. The ultimate goal of the so-called why-How Ladder, which is where designers utilize why to advance, is to discover how to address one or more issues. Your How Might We inquiries will assist you in progressing from the Define stage of Design Thinking into the subsequent Ideation stage, when you begin searching for particular creative answers. In other words, you might argue that the Why-How Laddering begins with a Why question before figuring out How to address a particular issue or design difficulty.

Module 3

Ideation

3.1 Idea Generation

Ideation is a creative process where designers generate ideas in sessions (e.g., brainstorming, worst possible idea). It is the third stage in the Design Thinking process. Participants gather with open minds to produce as many ideas as they can to address a problem statement in a facilitated, judgment-free environment. It's challenging to gain the perspective to find design solutions. To have productive ideation sessions, you'll need a dedicated environment for standing back to seek and see every angle. First, though, your team must define the right problem to address. Ideation, or "Ideate", is the third step in the Design Thinking process - after "Empathize" (gaining user insights from research/observation) and "Define" (finding links/patterns within those insights to create a meaningful and workable problem statement or point of view).

Before starting to look for ideas, your team needs a clearly defined problem to tackle - a focused problem statement or point of view (POV) to inspire and guide everyone. "How might we...?" questions—e.g., "How might we design an app finding cheap hotels in safe neighborhoods?"—help in reframing issues and prompting effective collaboration towards potential solutions. To bring people together to conjure ideas and bypass established frontiers, you need a skilled facilitator and a creative environment, including a prepared space, featuring posters of personas, relevant information, etc. Your team also requires rules - e.g., a 2-hour time limit, quantity-over-quality focus, ban on distractions such as phones, and "There are no bad ideas" mindset. By being bold and curious, participants can challenge commonly held beliefs and explore possibilities past these obstacles. Team members should take each other's ideas and build on them, find ways to link concepts, recognize patterns and flip seemingly impossible notions over to reveal new insights.

There are hundreds of ideation techniques to help you in your ideation sessions. You want an ideation technique that combines your conscious and unconscious mind—fusing the rational with the creative. It must match the sorts of ideas your team must generate and reflect their nature, needs and experience with ideation. Some crucial ones are:

Brainstorming – You build good ideas from each other's wild ideas.

Brain dumping – This is like brainstorming but done individually.

Brainwriting – This is like brainstorming, but everyone writes down and passes ideas for others to add to before discussing these.

Brain walking – This is like brainwriting, but members walk about the room, adding to others' ideas.

Worst Possible Idea – You take an inverted brainstorming approach, emboldening more reserved individuals to produce bad ideas and yielding valuable threads.

Challenging Assumptions – You overturn established beliefs about problems, revealing fresh perspectives.

Mind mapping – You use this graphical technique to connect ideas to problems' major and minor qualities.

Sketching/Sketch storming – You use rough sketches/diagrams to express ideas/potential solutions and explore the design space.

Storyboarding – You develop a visual problem/design/solution-related story to illustrate a situation's dynamics.

SCAMPER – You question problems through action verbs ("Substitute", "Combine", "Adapt", "Modify", "Put to another use", "Eliminate", "Reverse") to produce solutions.
Bodystorming – You use role-playing in scenarios/customer-journey steps to find solutions.

Analogies – You draw comparisons to communicate ideas better.

Provocation – You use an extreme lateral-thinking technique to challenge established beliefs and explore paths beyond.

Movement – You take a "what if?" approach to overcoming obstacles in ideation and finding themes/trends/attributes towards reliable solutions.

Cheat storm – You use previously ideated material as stimuli.

Crowd storming – Your target audiences generate and validate ideas through feedback (e.g., social media) to provide valuable solution insights.

Creative Pause – You take time to pull back from obstacles.
Other methods for ideation include co-creation workshops (combining user empathy research, ideation and prototyping), game storming (gamification-oriented ideation methods) and prototyping. The beauty of ideation is its unbounded freedom, although structured environments are critical. If you get stuck, you have fallbacks: e.g., "breaking the law" (listing constraints to see if you can overcome them), "stealing" ideas (emulating applicable concepts from other industries), inverting the problem and laddering (moving problems between the abstract and the concrete).

The process of creating, developing, and communicating ideas which are abstract, concrete, or visual. The process includes the process of constructing through the idea, innovating the concept, developing the process, and bringing the concept to reality.

Ideas are the key to innovation. Without them, there isn't much to execute and because execution is the key to learning, new ideas are necessary for making any kind of improvement.

It's obvious that ideas alone won't make innovation happen, as you need to be able to build a systematic process for managing those ideas. The point of ideation isn't just about generating tons of them but about paying attention to the quality of those as well.

We acknowledge that it can be challenging to generate more of those excellent ideas at times. It's not unusual to get stuck in our old habits and routines when we're supposed to be creating something new.

To break out of that rut, we've decided to look into idea generation as a whole as well as to introduce some of our favorite **tips, tools and**

techniques that can be used to generate new ideas more systematically.

Managing Ideas

Although generating ideas is often the easy part of innovation, collecting and managing them can be challenging without a proper tool.

Because people often come up with new ideas then and there, they should be able collect these ideas right when they arise.
Many people use digital or physical notebooks to write down their ideas on the go. This method works well when you just want to make notes for your own use. However, when you want to share that information with others and ask for feedback for further improvement, this isn't the most effective way.

If you need to involve more people in your ideation process, you might want to consider a dedicated idea management tool to make sure all of the potential ideas are collected in one place.

Idea Generation Steps

Idea generation follows a systematic approach that is often supported by an innovation process or a continuous improvement process. More and more companies are creating a culture of innovation and introducing continuous improvement managers to support employees in developing their ideas. This process initially includes a step in which a profound understanding of the task is gained. This can be done, for example, through customer interviews or a trend analysis. Questions are developed from this first step in idea generation. In the second step, ideas are generated. In the third step, the concepts are optimized and an implementation plan is drawn up.

One such method of systematic idea generation breaks down the development methodology of the well-known inventor Thomas Alva Edison into comprehensible steps and makes them applicable. In six consecutive steps, you can develop very effective ideas and create concepts that are ready for implementation.

Thomas Edison usually took the following steps:

Enabling: The search for the right field of innovation
Defining: Develop search queries and specify search paths
Inspiring: Search for thoughts and stimuli from other areas
Selecting: Generate and evaluate ideas
Optimizing: From the initial idea to the mature concept
Nurturing: Enrich ideas with various implementation strategies

3.2 Managing Innovation and Entrepreneurship

'An innovation is an idea, practice or object that is perceived as new by an individual or other unit of adoption".

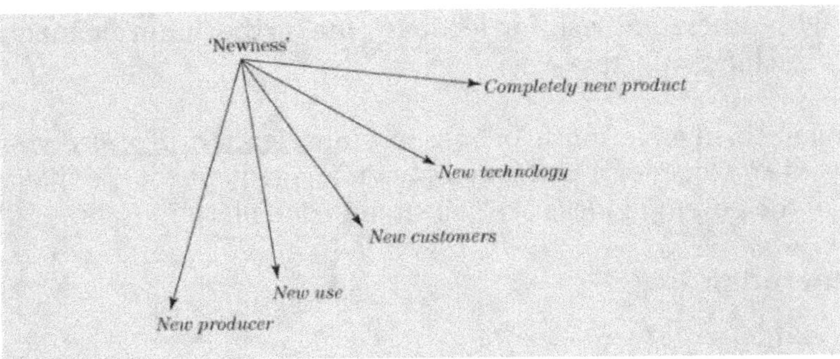

Figure 3.1 Newness in innovation

So, if innovation is something that attracts a lot of attention and that lots of people are interested in and regard as important – what exactly is it? Newness and novelty are key aspects of innovation, indeed the term is derived from the Latin word novus, meaning new or novel. However, as Figure 3.1 shows, 'newness' can take many different forms ranging from a completely new product to new uses for an existing product or simply a product that is new for a firm. This is captured in Rogers' (2003) definition of innovation which refers to differences in 'newness':

Innovation is a word deriving from Latin and means the introduction of something new to the existing world and the order of things or the improvement of resources productivity as mentioned by J. B. Say, quoted in Drucker (Drucker 1985). Many definitions of innovation are found in the literature. We report some of them: Chris Freeman and Soete (1982) reports: "The industrial innovation involves technical design, manufacturing, administrative and commercial activities related to the marketing of few (or improved) products or with the first commercial use of a new (or improved) process or equipment". Paul Gardiner (1985) highlights the following: innovation does not only mean commercialization of a significant advantage at the highest technical level (radical innovation), but it also includes taking advantage of small scale changes in the know-how (improvement or incremental innovation)..." Peter Drucker (1985) stresses that: "innovation is the special tool of businessmen to utilize change as an opportunity for a different activity or service. It is possible to appear as a discipline, to be learned, to be practiced". Paul Michael Porter (1990): "enterprises acquire a competitive advantage through acts of innovation. They approach innovation in its broader sense, including new technologies and the new way to do things". The term Innovation may refer to the process-conversion of an idea into a merchandised product or service, a new form of business organization, a new or improved functional production method, a new product presentation way (design, marketing) or even to a new service rendering method. It may also refer to the design and construction of new industrial equipment, the implementation of a project with a new management or may refer to a new way of thinking to deal with a situation or a problem. (Green Paper of the E.U. on innovation). Technological evolution and the parallel social and economic changes take place through the realization of innovation. A society's ability to innovate largely constitutes a mechanism of renewal and development. Innovation regards every aspect of economic or productive process.

At the level of an enterprise or an organization, innovation is mainly realized either by developing new products and services or by restructuring production–operation processes. The continuous innovative effort for new products-services or new productive processes create a competitive advantage in three critical areas:

a. Evaluation of the resources involving research and development activities, application of a new technology, sales productivity, production etc., new productive investments and expansion into new markets or broadening of the customer base.

b. Development and renewal of the entity with investments and growth, professional evolution opportunities for human resources, new recruitments and optimism, high morale and spirit.
c. Business success building on the reputation and attracting new customers, image of a dynamic business, products that distinguish from the competition, ongoing development and making hard for the competition to gather pace.

Innovation is often associated with the creation of a sustainable market around the launching of a new and superior product or process. In the literature of technology management, technological innovation is characterized by the introduction of a new technological product in the market: 'Technological innovation is defined here as a situationally new development through which people extend their control over the environment. Essentially, technology is a tool of some kind that allows an individual to do something new. So, technology transfer amounts to communication of information, usually from one organization to another'. (Tornatzky and Fleischer 1990) In particular, technological innovation is defined as: 'Introduction in the market of a technologically new or significantly improved product or the application of a technologically new or significantly improved productive process, successfully responding to market demand. It is the outcome of the interplay of market conditions on the one hand and of the possibilities to utilize the stock of technological and scientific knowledge' (Schumpeter 1934) Many authors acknowledge the importance of technological innovation for a company's high performance today.

We report some of them:

- Technological innovation in various enterprises is one of the key reasons for industrial competitiveness and national development (Freeman and Soete 1982; Porter 1985).
- Innovation is the only special ability in the 1990s (Peters 1996).
- The key feature of a modern market is not the price, but innovation (Zaltman et al. 1973).
- Innovation is the ultimate border in the modern business world helping companies to attain better yield, new products and services at lower cost (Pospiril 1996).

Technological innovation is a new technology that creates new products, hence new opportunities for the industry. This is the basic meaning of innovation and the reason it is important for economic growth as it creates business opportunities. Technology was and will be the key incentive to drive changes in our society. Technological innovation has turned into the largest driving force paving the way for society since 1980s. There is a constant flow of products and processes, from power engines in cars, airplanes, telecommunications and pharmaceutical preparations. All enterprises owe their existence and long-standing presence to the successful application of technology in the creation of new products and improved manufacturing processes. At present, management executives acknowledge the primordial role of technological innovation in a company's business success (Hans J. Thamhain). The enterprises that failed to keep up their innovative nature have been outstripped by younger and more active organizations. This failure is due to their weakness to see through the impact of new technology, while competitors seize the opportunity for development offered by technology.

Traditionally, innovation has been linked to R&D activities and the use of technological knowledge. This could be explained by the fact that all examples outlined in the literature of businesses are taken from the pharmaceutical industry, the chemical industry or electronics, where basic research has brought about the innovation that changed the world (penicillin, nylon and microprocessors). Any industrial application of scientific knowledge is a technological innovation. It should be stressed at this point that all sectors, of high or low technology, can use technology for innovation.

From a company's perspective, innovation is considered as the fortunate ending of an invention's commercialization journey and results in a constantly thriving market share or a new market. In other words, a technical discovery or invention (or creation of something new) is not important for a company unless this new technology can be utilized to add value to the company through income increase, cost cutting and similar improvements in economic results. This has two important consequences for the analysis of any innovation in a business organization. Firstly, innovation should be in-built in the organization's functions and strategy to clearly impact on the way said organization creates value or on the type of value offered in the market by this organization. Secondly, innovation is a social process, as it is only through intervention and management of persons that an organization can actualize the benefits of innovation. The discourse around innovation leads clearly to the creation of a model for the understanding and the evolutionary nature of innovation. Innovation

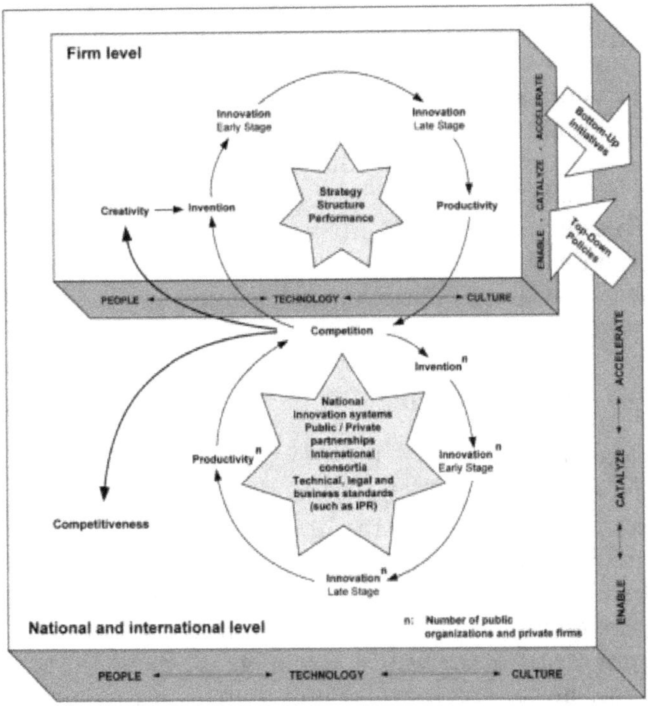

Figure. 3.2 Creativity, innovation, competitiveness

management deals with the activities undertaken by the enterprise to produce solutions for problems related to products, processes and management. Innovation involves uncertainty and dis-equilibrium. Nelson and Winter (1982) suggested that any change-however insignificant-represents an innovation. They also suggest that **figure. 3.2** Creativity, innovation, competitiveness considering uncertainty, innovation ends up to the emergence of new technologies and changes in relative weighting of existing technologies (ibid). This results in the disruptive process of dis-equilibrium. Insofar as innovation is adopted and disseminated, existing technologies become less useful (reduction of weight factors) or even useless (weight equal to "0") and are discarded. The stage of adoption is when uncertainty appears. New technologies are not adopted automatically but rather the markets influence the adoption rate (Carayannis and Alexander 1997; Carayannis and Alexander 1998a, b; Carayannis and Jorge 1998c).

There are several recurrent basic principles pertaining to innovations. These principles are summarized as follows:

- The integrated organizational approach
- The incentives of innovations
- The systematic process to convert an invention into innovation
- Team skills
- Communications
- Learning and
- Project management.

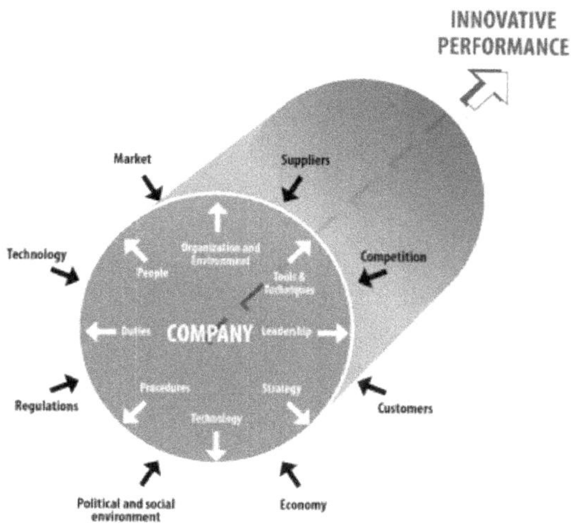

Figure 3.3 Factors influencing innovative performance

We believe that competitiveness is a product and a function of creativity and innovation reserves and supply, being determined and modified through various types, ways and kinds of knowledge (up to bottom and bottom up, acting proportionately, succeeding or failing through the exchange of capabilities, cooperatives, technological activity, supranational knowledge, domestic knowledge as well as through special knowledge and inventiveness (the 'when', 'how' and 'why' of creativity and innovation).

The innovation process should be guided by the clearly delimited market focus and involves 4 basic activities:

1. Creation of new Ideas:

    - Determination of new ideas of products and services.
    - Customer needs forecast with the analysis of market trends and competitors' successes.
    - Encouragement of new ideas and creativity among the personnel.
    - Determination of the mechanisms and the criteria used for the selection of ideas being developed.
    - Programming the creation of new products.

2. Re-design of production processes:

    - Re-design of production processes to achieve higher flexibility or productivity, along with higher quality and reduced production costs.
    - Changes in production processes to allow changes in the products.
    - Evaluation of the introduction of new technologies and management of organizational tools in production processes to increase the value of products.

3. Knowledge Management:

    - Innovation because of technology and know-how.
    - The way and the procedures whereby enterprises decide what technologies to develop internally (current training, set up of an R & D department, etc.).
    - The extent to which enterprises are supplied with technology or make available technology in external centers, as it happens with the subcontracting of Research & Development programs to other businesses, research centers and universities.

4. Product development:

    - How are enterprises driven from an idea to the launching of a new product or service in the market? This process requires a detailed description of operations and product specifications, such as the description of the parts and the systems it contains, and it should be taken into consideration in the production and distribution process as well as in the product's after-sale services.
    - How do enterprises develop a new product as soon as possible?
    - How do enterprises coordinate the internal personnel and the external teams?
    - What kind of management methods shall be applied by enterprises?

5. Re-design of market processes:

    - The changes in the marketing processes contribute to the increase of products' value or to the creation of new products and services.
    - Use of the modern information technology to redefine the market's product.

Managing Innovation is the entire process by which an organization generates creative new technological ideas (invention) and converts them into novel, useful and viable commercial products, services, and business practices for (potential) economic gain

- Analyze the key elements that together make entrepreneurial organizations more effective at technological innovation
- Build on ideas that you have learned in other classes (e.g. teams, leadership, networks etc.)
- Serve as the foundation for other courses on innovation & entrepreneurship
- Provide "action-relevant" knowledge & skills to go into a firm and diagnose its innovation challenges & to organize your own innovative projects

The opportunity and challenge for entrepreneurs around the world to accelerate and affect economic growth and leverage the Digital Divide through business initiatives in the private sector. As said by innovation is related to shifting resources to areas of higher yield. Therefore, knowledge-based and knowledge-supported entrepreneurship will be the pre-eminent driver of innovation in the twenty-first century, via real/virtual and global/local infrastructures such as the incubator networks listed below. This vision is particularly promising and appealing in the context of e-Development towards the Knowledge Economy.

The Entrepreneur, therefore, with the help of a decision-making grid, takes the following actions:

- Makes decisions and takes risks
- Organizes factors of production in the most efficient way aiming to achieve the best possible result
- Uses innovations and monitors pioneering entrepreneurial initiatives in order to draw ideas
- Seeks profit opportunities and adopts clever strategies.

Internal Entrepreneurship: The term Internal Entrepreneurship is used to describe the existence of inherent entrepreneurial tendencies of big company employees. Internal Entrepreneurship can be cultivated by rendering the senior staff personally responsible either for the preparation of a budget corresponding to their tasks, or for managing their departments as independent self-managed companies, regarding budget preparation and control. Internal Entrepreneurship focuses on initiative and spirit of enterprise: an executive possessing internal entrepreneurship behaves as if he owns the company. Consequently, internal entrepreneurship is a cognitive state which exhibits the following distinct characteristics:

- Persistence and determination manifesting themselves in the desire for very hard work
- Willingness to take risks but only after careful research about the chances of success
- Understanding, mainly based on how the market functions, particularly in relation to identifying new opportunities

- Reluctance to be trapped in conventionality and the gears of bureaucracy, with an aversion to restrictive regulations

- Enthusiasm for exercising entrepreneurial activities, emotional commitment to an organization, ability to inspire confidence in subordinates, and self-confidence to external suppliers, customers and other third parties.

External Entrepreneurship: External entrepreneurship includes elements of external business environment affecting a company, either directly or indirectly. External entrepreneurship is divided into general and specific. General entrepreneurship deals with sizes and forces which somehow affect the life and work of the agencies mentioned above. These elements, however, cannot be specifically identified, nor can the degree of their influence on the company be assessed with certainty and precision. One main characteristic of general entrepreneurship is that the situations prevailing and the driving forces developing in its context affect each and every business.

3.3 Exploring Innovations – Processes and Methods

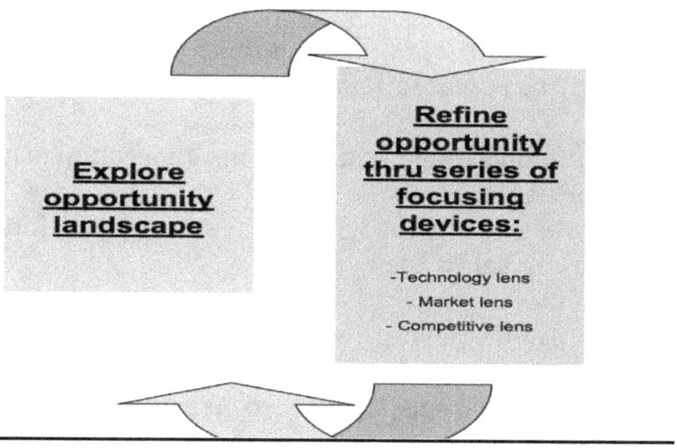

Figure 3.4 Six tools to improve your creativity during an innovation process

Creativity is not a gift but a skill we can train with proper tools. When we introduce somebody to innovation methodologies, especially those focused on business models, the role of creativity topic usually comes up. The most common question is whether we need to use creativity or not to innovate. There is a short answer for this: Most of the successful innovation methodologies give creativity a strong role. Moreover, all those methodologies clearly state that creativity is not a matter of genius or lucky lightbulb moments, but something that everyone can achieve with the help of proper tools.

From this basic statement we can go further into the topic of creativity in innovation. We can explain that the tools for creativity are used during a regular innovation process where teams should generate new ideas to further explore their feasibility as proper business models. For individuals not accustomed to innovation methodologies (which is almost everyone), the phases that utilize creativity are the most challenging.

Because of this challenge, innovators have developed a wide range of tools to help people unleash and nurture their innate creativity, which is usually constrained by the rules, procedures, and methods that many companies use. These restrictions are suited to run established companies, but they are not helpful to explore new possibilities during an innovation workshop or project.

Let us explore the most popular creativity tools for innovation. To get started it may be useful to separate them into two groups:

1. Divergence tools, focused on the generation of as many ideas as possible.
2. Convergence tools, which strive to analyze, filter and merge ideas, to select the best ones.

Within both groups there are a wide range of options to choose from, depending on the workshop and project. Below we will explain some of the most common tools, which also give us a short glimpse of what creativity for innovation means.

Divergent Tools

All the creativity tools focused on divergence try to generate as many ideas as possible during a short time span, usually a few minutes during an ideation session. A basic rule here is not to criticize or judge new ideas as they come, as negativity is probably the best way to kill any ideation. Innovators should let their brains run free to generate as many possibilities as they can; filtering and merging will come later, during the convergence process.

Analogy Cards

BMI Pattern Cards is a good example for the kind of analogy cards used to improve creativity during an innovation process. Analogy thinking produces 80% of ideas during the ideation session. Basically, this means copying concepts from other industries and adapting them to our own. So, a great way to generate new business ideas is to check and review other business models and industries, even those not closely related to our own company. Try to analyze their business model and Figure out how we can apply it to our industry. During this process, new business possibilities will naturally arise.

Analogy cards, such as BMI´s Business Pattern Cards, are a great way to practice analogy thinking, since they give a clear reference to work while guiding the creativity process for better results.

Trend & Technology Benchmarking

Another alternative method for analogy thinking uses technology benchmarking as a reference. Sometimes it is useful to select several trend technologies, and try to Figure out how to implement them in our company. For instance, if we choose IA, IoT, and blockchain, through a brainstorming process we would come up with new business ideas for our company based on current applications for these technologies. As we do that, we will find that new ideas come from the interlink between our experience and new possibilities coming from new technologies.

The "three ifs"

Lateral thinking tools are another useful methodology set to run an ideation session. This way of thinking involves looking at your situation in a different way -- essentially stepping sideways from our current business strategy. A nice way to do that is the "three ifs" tool, which is very useful to challenge our assumptions and change our perspective. In this phase, we should ask ourselves what would happen if our situation was different in three ways. For instance:

1. What would happen if I change my client?
2. What would happen if I have another value proposition?
3. What would happen if I change my value chain?

For better results, it is recommended to combine analogy with lateral thinking. For example, we can ask ourselves what would happen if we introduce AI to our value proposition. Or maybe we can try to Figure out which clients we would have if we applied a particular business model from another industry to our own. While trying to answer these "if" questions, we will find that our previous business knowledge will help us to discover new possibilities, generating new ideas during the process. The possibilities for analogy and lateral thinking are huge, and that is fine, since these tools are designed to fire the spark for a fruitful ideation process, where we can widen our business scope.

Convergence Tools

Convergence thinking tools are intended for a clear purpose: to help us to organize, filter and select all the ideas generated during the divergence process. This step requires a very different mindset, so the tools used for this purpose are also clearly different, focusing on analysis and synthesis, rather than idea generation. Many people might think that this kind of work is not creative, but in fact, it is. If we want to successfully finish any innovation process, running only the ideation step is useless. We should also have a clear method to filter the best ideas, leaving apart those not worthy enough, and focusing on the best ones.

Mind Maps

Business model canvas is a great way to gather, organize and filter information and ideas coming from the ideation process.

Mind maps are usually a great way to organize ideas. We can use them to explore associations and divergences between ideas, clustering all those concepts coming from ideation. Mind maps will help to identify underlying trends, cleaning redundancies and narrowing our scope to fewer concepts to work with. Usually, we need to create a mind map to achieve this, although we can also use several prebuilt canvases and tools, such as a customer and business model canvas.

Thinking Hat

The thinking Hat is a popular technique, very useful to explore the upsides and downsides of every idea. We might define several hats, each one representing a different mindset (for instance: the customer, the advocate, the devil´s advocate, and the board of directors). After pitching the idea, some people should wear different hats, analyzing the pitch from a different perspective. It might be useful to have the same person wear several hats, forcing him or her to look at the same idea from different angles. The feedback this technique generates is valuable and helps the team to bring new perspectives into the discussion, dismissing those ideas that do not make it through the test, and improving those passing it.

Hundred-Euro Test, or Idea Shopping

This tool is very useful for rating ideas. We can introduce several value propositions to a selected audience, usually in a canvas format, asking people to buy whatever proposition they want with just 100 euros. Euros can be freely allocated and expended in several places or in just one. No prices are set, so everybody gives to a particular idea whatever they think is worth paying. After the funds are distributed, we can easily identify the most popular ideas, as they will receive most of the money. This test allows the team to get valuable feedback from outsiders, filtering ideas and leaving aside prejudices.

Creativity as a Circular Process

Both kinds of tools, divergent and convergent, are used during different phases of innovation. Sometimes we need to expand our scope to have a wider view of the choices we can make. After that, probably we will need to focus on a few ones, just to move forward in our exploration of new ideas and business models. For both purposes, we should choose the proper tools.

Moreover, this innovation process is not a linear path. We can go through the divergence and convergence process as much as we need, looking for new and potentially feasible ideas. We must also keep in mind that creativity is about teamwork. We should encourage constructive criticism, including all perspectives on the process. Diverse teams are usually much better than homogeneous ones, since different mindsets and experiences can merge and enhance each other.

Depending on our goals, we might also create a strategy that mixes both tools, sometimes settling with one tool is not easy and requires a facilitator with experience and patience to lead the process, choose the right tools and lead people down the same path. With professional help, we will make the most of the tools chosen, unlocking the potential of any team.

3.4 Understanding the Discovery Process for Opportunities

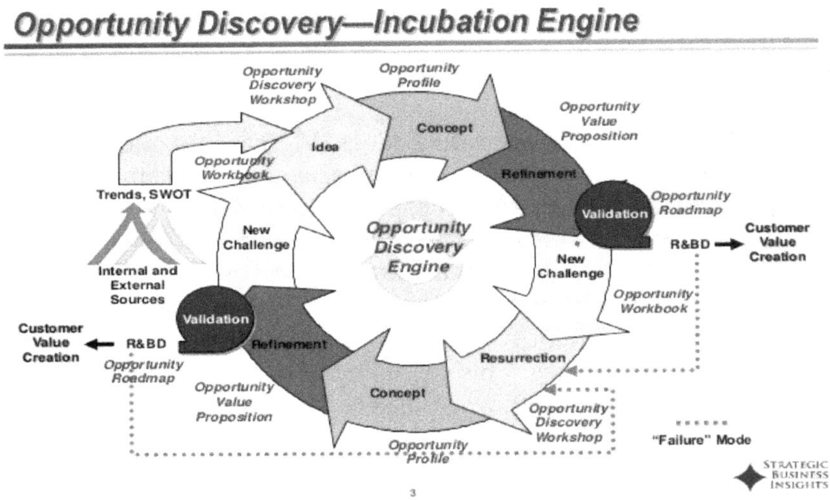

Figure 3.5 Opportunity discovery-incubation engine

Invest time in discovery. Generally, the more time you spend on a discovery, the more details you will find about a problem you solve, the better solution you will find.

There are two types of discovery conducted independently: feature/functional discovery and design discovery.

Feature/Functional Discovery

Usually, this discovery is conducted by UX analytics. The goal of feature and functional discovery is to collect all requirements for a future product, research the current state of a market to find the competitors, and establish a technological stack for a future product (choose technologies that can be used in this product). A team involved in feature discovery does the following things:

Analyze User Expectations

What users need and what users want are not the same things. It's too often product team focus on both needs and wants, resulting in an infinite product backlog. You need to know who your target users are and what they need. You can gain this knowledge by interviewing people within the target audience.

Clearly Define Business Objectives

When it comes to product design, we have two primary goals – create a great user experience for people who will use our products and satisfy business objectives. The goal of feature discovery is to align the business objectives and product design.

Create a Competitive Comparison

After a team conducts a competitor's analysis, it'll have a solid understanding of the direct and indirect competitors. The next step is to use this information to understand what the strong and weak sides of your future product will be. This information can be presented to stakeholders in a visual way – you can create a chart called Competitive Comparison. This chart will make it clear both to you and your stakeholders to what niche your product belongs.

Design Discovery

Usually, this discovery conducted by designers. Designers research what design decisions are applicable to a particular product.

A crucial moment during the design discovery is not to be restricted by limitations. Each limitation will have a negative impact on creativity – it will prevent genuinely innovative products. Only when you free your mind, you can come up with a genuinely creative design decision.

The goal during the discovery stage is to connect the information from both discovery processes together and create a robust product design direction. After both teams finalize their activities, they should meet and discuss their design decisions. This is the moment to merge the design decisions and establish a solid direction for design. It's vital to stress a few important moments:

Designers should never blindly follow the engineering team. The design direction should be reached based on a mutual agreement, but not forced by the engineering team.

Identify opportunity. During the discovery phase, it's vital to understand not only the current state of the market but also to project this state to the future. It will help you design answers to where the product will be in two, five, or ten years. This will increase the probability that a product will be in demand and sustainable.

Here is a quick practical example. Suppose your team works on a human-machine interaction (HMI) for a premium car.
The first team collects all requirements from stakeholders. Then it conducts an analysis of premium car brands, the technologies they use, and the solutions they provide for their customers. The team analyses both benefits and downsides of different solutions. Based on this information, the team proposes a set of features that should be implemented in a product.

The second team does design discovery. They don't focus on functions and technologies, they focus on visual decisions that can be implemented in a project. For example, a team might analyze the shape of a car to find the perfect curve for an HMI.

Ideation Phase

When it comes to product design, we often hear the word "roadmap." Many product teams believe that a roadmap is something that can be created before starting a product development and this roadmap prevents them from making mistakes (a.k.a. incorrect product decisions). In reality, it's almost impossible to create a product roadmap right at the beginning of a project. Building a solid roadmap might take months or years (if we talk about sophisticated digital products). But the good news is that proper planning makes this process more efficient. The goal of designers is to set a path – form a product strategy and establish the right direction for a product. Part of it can be done during the discovery phase and the other part is during the ideation phase.

During the ideation phase, it's vital to have strong brainstorming sessions. Follow these tips on how to brainstorm more efficiently:

Ask the most creative people who are involved in a project to join the session. The session itself should take place in a physical space, not digital space, if possible. Physical spaces make it easier to share emotions. Each session participant will not only be able to share their thoughts but also see the physical reaction of other people.
Dream big. When it comes to ideas, you should not limit yourself. Relax and imagine what a perfect product would look like. If you work on a cutting-edge product, think about some device you saw in sci-fi films.

A scene from Blade Runner (1982). The technologies we saw in sci-fi movies were a dream for a long time. But today we can have many of them.

Structure your ideas. Without structuring the ideas you collect, it's fairly easy to lose yourself in that stream. We use boards with stickers. Each idea is placed on a sticker and belongs to one group. Our team uses the following groups: Emotional Needs, Practical, Safety, Security, Entertainment, Media, Navigation, etc. The color of a sticker helps us to group the ideas.

Consider the technical feasibility. Not all ideas can be implemented using modern technologies. It's essential to discuss your ideas with an engineering team.

Tips for Ideation Phase

Always keep the big picture in mind. The primary goal of the ideation phase is not to find a simple solution to a particular design challenge. The goal is to establish a user experience strategy.

When it comes to product design, always have the answer to the question, "Why do I need to do it?" Designers should not create something just because they can do it. Having a clear answer to why will help prevent extra work and save you a lot of hours.

Be inspired. Inspiration plays a vital role during the ideation phase of the process. It's clear that without inspiration it's really hard to create something outstanding. Designers have a few powerful tools in their toolboxes that allow them to collect and analyze great design examples. I just want to share one tool here – mood boards. Mood boards bring two advantages to product designers:

It helps you find something beautiful that you have in a world to implement in your design. For example, you can take inspiration from nature and use the colors from a particular photo in your mobile design product.

Mood boards often help designers to explain and illustrate "why" they make a particular design decision to peers and stakeholders. All too often designers have a specific design problem they want to solve, and a mood board allows them to find a proper solution for this problem. When it comes to the question, "Why do we need this specific shape in our UI?" you can show the mood board and your design decision will become more clear to them.

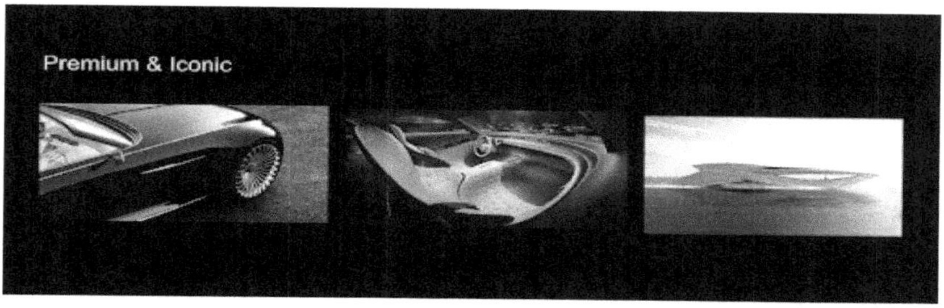

After combining together around 1000 images, I leave only 10 of the most powerful and consistent visual references on my moodboard. Moodboard created by Gleb Kuznetsov

Convey emotion in your design. All too often we focus on the usability of our products but we forget about the emotional aspects of interaction. Emotions are equally important for product design because the emotional interfaces can better connect with users

3.5 Process for identifying Opportunities - Brainstorming

What Is Brainstorming?

Madison Avenue advertising executive Alex Osborn developed the original approach and published it in his 1953 book, "**Applied Imagination**." Since then, researchers have made many improvements to his original technique.

The approach described here takes this **research** into account, so it's subtly different from Osborn's approach.

Brainstorming combines a relaxed, informal approach to problem solving with lateral thinking. It encourages people to come up with thoughts and ideas that can, at first, seem a bit crazy. Some of these ideas can be crafted into original, creative solutions to a problem, while others can spark even more ideas. This helps to get people unstuck by "jolting" them out of their normal ways of thinking.

Therefore, during brainstorming sessions, people should avoid criticizing or rewarding ideas. You're trying to open up possibilities and break down incorrect assumptions about the problem's limits. Judgment and analysis at this stage stunts idea generation and limit creativity.

Evaluate ideas at the end of the session – this is the time to explore solutions further, using conventional approaches.

Why Use Brainstorming?

Conventional group problem solving can often be undermined by **unhelpful group behavior**. And while it's important to start with a **structured, analytical process** when solving problems, this can lead a group to develop limited and unimaginative ideas.

By contrast, brainstorming provides a free and open environment that encourages everyone to participate. Quirky ideas are welcomed and built upon, and all participants are encouraged to contribute fully, helping them develop a rich array of creative solutions.

When used during problem solving, brainstorming brings team members' diverse experience into play. It increases the richness of ideas explored, which means that you can often find better solutions to the problems that you face.

It can also help you get buy-in from team members for the solution chosen – after all, they're likely to be more committed to an approach if they were involved in developing it. What's more, because brainstorming is fun, it helps team members bond, as they solve problems in a positive, rewarding environment.

While brainstorming can be effective, it's important to approach it with an open mind and a spirit of non-judgment. If you don't do this, people "clam up," the number and quality of ideas plummets, and morale can suffer.

Individual Brainstorming

While group brainstorming is often more effective at generating ideas than normal group problem solving, several **studies** have shown that individual brainstorming produces more – and often better – ideas than group brainstorming.

This can occur because groups aren't always strict in following the rules of brainstorming, and bad behaviors creep in. Mostly, though, this happens because people pay so much attention to other people that they don't generate ideas of their own – or they forget these ideas while they wait for their turn to speak. This is called "blocking." When you brainstorm on your own, you don't have to worry about other people's egos or opinions, and you can be freer and more creative. For example, you might find that an idea you'd hesitate to bring up in a group develops into something special when you explore it on your own.

However, you may not develop ideas as fully when you're on your own, because you don't have the wider experience of other group members to draw on.

Tip: To get the most out of your individual brainstorming session, choose a comfortable place to sit and think. Minimize **distractions** so that you can focus on the problem at hand, and consider using **Mind Maps** to arrange and develop ideas.

Individual brainstorming is most effective when you need to solve a simple problem, generate a list of ideas, or focus on a broad issue. Group brainstorming is often more effective for solving complex problems.

Group Brainstorming

Here, you can take advantage of the full experience and creativity of all team members. When one member gets stuck with an idea, another member's creativity and experience can take the idea to the next stage. You can develop ideas in greater depth with group brainstorming than you can with individual brainstorming.

Another advantage of group brainstorming is that it helps everyone feel that they've contributed to the solution, and it reminds people that others have creative ideas to offer. It's also fun, so it can be great for team building!

Group brainstorming can be risky for individuals. Unusual suggestions may appear to lack value at first sight – this is where you need to chair sessions tightly, so that the group doesn't crush these ideas and stifle creativity.

Where possible, participants should come from a wide range of disciplines. This cross-section of experience can make the session more creative. However, don't make the group too big: as with other types of teamwork, groups of five to seven people are usually most effective.

How to Use the Tool

You often get the best results by combining individual and group brainstorming, and by managing the process according to the "rules" below. By doing this, you can get people to focus on the issue without interruption, you maximize the number of ideas that you can generate, and you get that great feeling of team bonding that comes with a well-run brainstorming session!

To run a group brainstorming session effectively, follow these steps.

Step 1: Prepare the Group

First, set up a **comfortable meeting environment** for the session. Make sure that the room is well-lit and that you have the tools, resources, and refreshments that you need.

How much information or preparation does your team need in order to brainstorm solutions to your problem? Remember that prep is important, but too much can limit - or even destroy - the freewheeling nature of a brainstorming session.

Consider who will attend the meeting. A room full of like-minded people won't generate as many creative ideas as a diverse group, so try to include people from a wide range of disciplines and include people who have a variety of different thinking styles.

When everyone is gathered, appoint one person to record the ideas that come from the session. This person shouldn't necessarily be the team manager - it's hard to record and contribute at the same time. Post notes where everyone can see them, such as on flip charts or whiteboards; or use a computer with a data projector.

If people aren't used to working together, consider using an appropriate warm-up exercise, or an **icebreaker** .

Step 2: Present the Problem

Clearly define the problem that you want to solve and lay out any criteria that you must meet. Make it clear that that the meeting's objective is to generate as many ideas as possible.
Give people plenty of quiet time at the start of the session to write down as many of their own ideas as they can. Then, ask them to share their ideas, while giving everyone a fair opportunity to contribute.

Step 3: Guide the Discussion

Once everyone has shared their ideas, start a group discussion to develop other people's ideas, and use them to create new ideas. Building on others' ideas is one of the most valuable aspects of group brainstorming.

Encourage everyone to contribute and to develop ideas, including the quietest people, and discourage anyone from criticizing ideas.

As the group facilitator, you should share ideas if you have them, but spend your time and energy supporting your team and guiding the discussion. Stick to one conversation at a time, and refocus the group if people become sidetracked.

Although you're guiding the discussion, remember to let everyone have fun while brainstorming. Welcome creativity and encourage your team to come up with as many ideas as possible, regardless of whether they're practical or impractical. Use thought experiments such as **Provocation** or **Random Input** to generate some unexpected ideas.

Don't follow one train of thought for too long. Make sure that you generate a good number of different ideas and explore individual ideas in detail. If a team member needs to "tune out" to explore an idea alone, allow them the freedom to do this.

Also, if the brainstorming session is lengthy, take plenty of breaks so that people can continue to concentrate.

Taking Your Brainstorming Further

If you're not getting enough good quality ideas, try using the approaches below to increase the number of ideas that you generate:

- **The Stepladder Technique** - This improves the contribution of quieter group members by introducing one person at a time.

- **Brainwriting** - This is a written approach that you can use to encourage all individuals to generate and develop ideas.

- **Online Brainstorming (also known as Brain-netting)** - An electronic method of brainstorming, this uses a document stored on a central server, or on a Cloud-based system.

- **Crawford's Slip Writing Approach** - You can use this approach to get plenty of Ideas from all participants, and to get a view of each idea's popularity.

These techniques help you in specific situations:

- **Reverse Brainstorming** - This is used to improve a product or service.
- **Starbursting** - Starbursting helps you develop questions that you need to ask to evaluate a proposal.
- **Charette Procedure** - This helps you brainstorm with large groups of people. (Conventional brainstorming becomes increasingly ineffective when more than 10 or 12 people are involved.)
- **Round-Robin Brainstorming** - You can use this approach to get people to contribute ideas without being influenced by others.
- **Rolestorming** - This technique encourages group members to take on other people's identities while brainstorming, thereby reducing their inhibitions.

3.6 Alex Osborn's Original Rules of Brainstorming

Alex F. Osborn: Father of the Brainstorm

"It is easier to tone down a wild idea than to think up a new one."- Alex Osborn

That quote pretty much sums up Osborn's ideas on brainstorming and creative thinking. Brainstorming is method of thinking up solutions, ideas or new concepts. It can be a difficult process for many reasons: sometimes people are unwilling to suggest a solution for fear of criticism or the problem may just be a very difficult one, and one that no existing solutions exist for yet. Osborn's solution, as suggested by the above quote, is to think up as many ideas as possible regardless of how ridiculous they may seem at first. Since it is very unlikely to think up the perfect solution right off the bat, he recommends getting every idea out of your head and then go back to examine them afterwards. An idea that may have initially sounded off-the-wall may actually turn out to be a plausible idea with a little modification. Osborn's technique of deferred judgment increases the individual's synthesis capabilities by releasing the human mind from the analysis mode of thinking. Brainstorming is considered to be a group method of listing suggested ideas pertaining to a solution for a specific problem.

Creative thinking requires tools such as the brainstorm and the affinity diagram. Brainstorming is simply listing all ideas put forth by a group in response to a given problem or question. In 1939, a team led by advertising executive Alex Osborn coined the term "brainstorm." According to Osborn, "Brainstorm means using the *brain* to *storm* a creative problem and to do so "in commando fashion, each stormer audaciously attacking the same objective." Creativity is encouraged by not allowing ideas to be evaluated or discussed until everyone has run dry. Any and all ideas are considered legitimate and often the most far-fetched are the most fertile. Structured brainstorming produces numerous creative ideas about any given "central question". Done right, it taps the human brain's capacity for lateral thinking and free association.

Other rules for brainstorming according to Osborn include creating an environment where team members are not criticized for their ideas. Ideas can be evaluated after the brainstorming session but judgments during the process will only alienate team members. Also, after the idea generating process team members should try to combine and modify ideas.

The technique of brainstorming takes place in a panel format. The brainstorming panel is composed of a leader, recorder, and panel members. The leader is responsible for maintaining a rapid flow of ideas while the recorder lists all the ideas as they are presented. The size of the panel may vary but a range of 10 to 12 is usually optimum. If the group were to get to large there is a greater chance of members remaining in the background or that the members would not have enough time to express their ideas in a reasonable amount of time. These practices are essential to any brainstorming process, but many people have criticized Osborn's methods as incomplete. For instance he does not suggest that team members prepare for brainstorming sessions. Another downfall of Osborn's teaching is that he does not offer and tools or exercises for coming up with creative ideas in the first place. However limited his methods may be they are productive means of acquiring new and innovative solutions.

SWOT Analysis SWOT analysis is another tool for auditing an organisation and its environment. It is the first stage of planning and helps markets to focus on key issues. A scan of the internal and external environment is an important part of the strategic planning process. Environmental factors internal to the firm can usually be classified as strengths (S) or weaknesses (W) and those external to the firm can be classified as opportunities (O) and threats (T). Such an analysis of the strategic environment is referred to as a SWOT analysis. SWOT analysis provides information that is helpful in matching the firm's resources and capabilities to the competitive environment in which it operates SWOT analysis is a business tool by which, a firm wishing to implement a strategic analysis, analyses and recognizes its corporate Strengths and Weaknesses as well as the existed or forthcoming Opportunities and Threats from its external environment. Only when these four critical information elements are well elaborated and known, the enterprise is able to formulate and implement the strategy leading to its business aims. The role of SWOT analysis is to take the information from the environmental analysis and separate it into internal issues (strengths and weaknesses) and external issues (opportunities and threats). Once this is completed, SWOT analysis determines if the information indicates something that will assist the firm in accomplishing its objectives (a strength or opportunity), or if it indicates an obstacle that must be overcome or minimized to achieve desired results (weakness or threat). When doing SWOT analysis, remember that the S and W are INTERNAL and the O and T are External. SWOT analysis is an extremely useful tool for understanding and decision making for all sorts of situations in business and organisations. SWOT Analysis is a very effective way of identifying your Strengths and Weaknesses, and of examining the Opportunities and Threats you face. Carrying out an analysis using the SWOT framework helps you to focus your activities on areas where you are strong and where the greatest opportunities lie. By creating a SWOT analysis, you can see all the important factors affecting your business together in one place. It's easy to create, easy to read, and easy to communicate. A company can use the SWOT analysis: • While developing a strategic plan or planning a solution to a problem. • In order to develop a plan that takes into consideration many different internal and external factors and maximizes the potential of the strengths and opportunities while minimizing the impact of the weaknesses and threats.

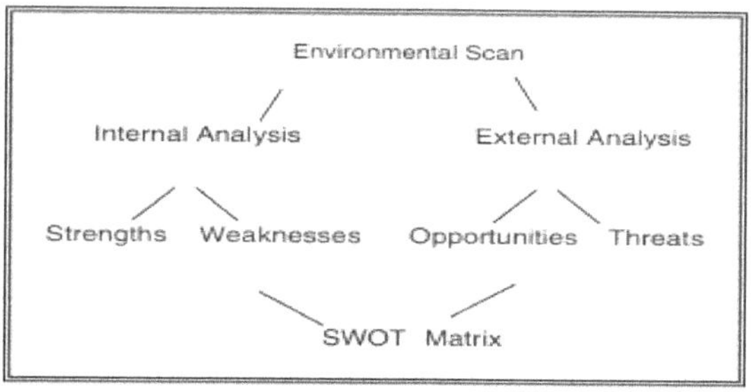

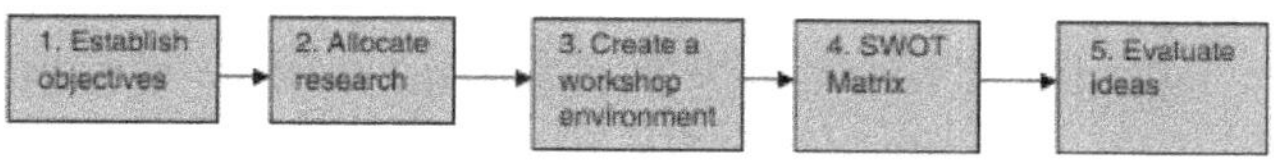

1. **Establish the objectives:** The first key step in any project is to be clear about what you are doing and why. The purpose of conducting SWOT analysis may be wide or narrow, general or specific.

2. **Allocate research and information-gathering tasks:** Background preparation is a vital stage for the subsequent analysis to be effective and should be divided among the SWOT participants. This preparation can be carried out in two stages:

- Exploratory, followed by data collection.

- Detailed, followed by a focused analysis. Gathering information on Strengths and Weaknesses should focus on the internal factors of skills, resources and assets, or lack of them. Gathering information on Opportunities and Threats should focus on the external factors.

3. **Create a workshop environment:** If compiling and recording the SWOT lists takes place in meetings, then do exploit the benefits of workshop sessions. Encourage an atmosphere conducive to the free flow of information and to participants saying what they feel to be appropriate, free from blame. The leader/facilitator has a key role and should allow time for free flow of thought, but not too much. Half an hour is often enough to spend on Strengths, for example, before moving on. It is important to be specific, evaluative and analytical at the stage of compiling and recording the SWOT lists.

4. List Strengths, Weaknesses, Opportunities, Threats in the SWOT Matrix.

5. Evaluate listed ideas against objectives: With the lists compiled, sort and group facts and ideas in relation to the objectives. It may be necessary for the SWOT participants to select their five most important items from the list in order to gain a wider view. Clarity of objectives is key to this process, as evaluation and elimination will be necessary to separate the wheat from the chaff. The SWOT analysis template is normally presented as a grid, comprising four sections, one for each of the SWOT headings: Strengths, Weaknesses, Opportunities, and Threats. The SWOT template below includes sample questions, whose answers are inserted into the relevant section of the SWOT grid. The questions are examples, or discussion points, and obviously can be altered depending on the subject of the SWOT analysis.

Strengths	**Weaknesses**
Advantages of proposition? Capabilities? Competitive advantages? USP's (unique selling points)? Resources, Assets, People? Experience, knowledge, data? Financial reserves, likely returns? Marketing—reach, distribution, awareness? Innovative aspects? Location and geographical? Price, value, quality? Accréditations, qualifications, certifications? Processes, systems, IT, communications? Cultural, attitudinal, behavioural? Management cover, succession?	Disadvantages of proposition? Gaps in capabilities? Lack of competitive strength? Reputation, presence and reach? Financials? Own known vulnerabilities? Timescales deadlines and pressures? Cash flow, start-up cash-drain? Continuity, supply chain robustness? Effects on core activities, distraction? Reliability of data, plan predictability? Morale, commitment, leadership? Accreditations, etc.? Processes and systems, etc.? Management cover, succession?
Opportunities	**Threats**
Market developments? Competitors' vulnerabilities? Industry or lifestyle trends? Technology development and innovation? Global influences? New markets, vertical, horizontal? Niche target markets? Geographical, export, import? New USP's? Tactics—surprise, major contracts, etc.? Business and product development? Information and research? Partnerships, agencies, distribution? Volumes, production, economies? Seasonal, weather, fashion influences?	Political effects? Legislative effects? Environmental effects? IT developments? Competitor intentions—various? Market demand? New technologies, services, ideas? Vital contracts and partners? Sustaining internal capabilities? Obstacles faced? Insurmountable weaknesses? Loss of key staff? Sustainable financial backing? Economy—home, abroad? Seasonality, weather effects?

- Tips for successful SWOT analysis

	Top tips	But remember …
1	Never copy an existing SWOT analysis; it will influence your thinking. Start with a fresh piece of paper every time	*You could use a standard template to help the ideas flow*
2	Set aside enough time to complete it	*You may need to come back to it several times before you are happy*
3	The SWOT analysis itself is NOT the result. It's only a tool to help you analyse your business	*Before you begin any analysis, you should know what you intend to do with the results*
4	A SWOT analysis is not a business school fad. It is a proven technique used throughout the business community	*You need to be comfortable working with it in your business*
5	Keep your SWOT analysis simple, readable, short and sharp	*It needs to make sense to outsiders (e.g. bank managers or investors) so don't use phrases or acronyms that only you understand*
6	Make sure you create an action plan based on your SWOT analysis	*You need to communicate this clearly to everyone involved*
7	A SWOT analysis only gives you insight at a single point in time	*You need to review it—probably quarterly—to see how the situation has changed*
8	Don't over-analyse. Try not to worry if it isn't perfect, just get the analysis done	*If you are going to act on the results, it needs to be accurate in all the important areas*

Module 4

Experimenting and Developing Concepts

4.1 Experimenting and Developing Concepts

A CD & E project is the concept work and associated experiments, using scientific methods and analysis, to address identified capability gaps. It comprises developing and validating a concept, using the CD&E methodology, within a planned time frame and a given resource allocation to find solutions to particular shortfalls.

The military problem to be solved is stated, hypothesis solution is prepared, tested, improved if needed, and culminates in a validated and approved concept, with propositions for its implementation.
A CD&E project includes:

- A deeper analysis of the problem and identification of possible conceptual solutions;
- The drafting of a concept;
- The development of solutions;
- The analysis of the adequacy of the solutions, and experimentation to refine and validate the proposed solutions.

In general, the CD&E project employs an iterative approach. 'Concept development', 'experimentation' and 'analysis' are the three main techniques used during a CD&E project. While concepts could be developed in many different ways, only the utilization of these three 2 techniques together guarantees the sound practice of analysis and qualitative assessment of conceptual outputs. The characteristic and the intensity of these individual elements can be different, dependent on the topic. None should be excluded from the beginning, and their use must be considered at each stage of the development cycles (spirals of development). The CD&E project covers the development of the concept until the implementation phase, where decisions are made with respect to acquisition and application of the proposed solutions.

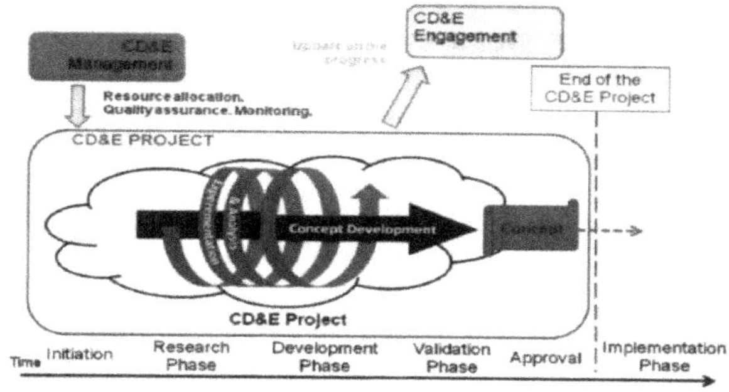

Figure 4.1 Concept development and experimentation project

A CD&E project is initiated, executed and controlled through accepted Project Management practices that permit timely decision, direction and guidance by senior leadership. Project Manager is to provide appropriate control mechanisms, forms, templates and methods of work to ensure a disciplined, transparent, structured and rational development and to enable inherent evaluation with efficient and well-defined validation and review points.

Project Initiation

The decision to apply CD&E to address capability shortfalls lays outsider the CD&E process and is made after a thorough analysis of the capability shortfall wherein the military problem is clearly stated. The decision process encompasses a stakeholder analysis, problem identification and structuring and a brainstorming of ideas. Once that decision has been made, a CD&E project is initiated. It will formally be part of the collection of CD&E projects managed by Project Manager.

Project Development

A typical CD&E project will display a concurrent, spiral process in which conceptual work, experimental efforts and analytical insights are intermingled to form a process that employs all skills from drafting of documents, brainstorming, wargaming, research and analysis, Modelling and Simulation (M&S), live experiments, and prototyping to demonstrations.

Thus development, refinement, assessment and validation are founded on a spiral development. Each specific project is entrusted to a Concept Development Team. It is composed, as appropriate, of concept developers, experimenters, analysts, M&S experts and Subject Matters Experts (SMEs). Other Experts interested in the project could also be a member of the Concept Development Team.

The Concept Development Leader leads the Concept Development Team and provides overall coordination and management of the CD&E project throughout its duration. He is responsible for producing the concept development plan and for providing information on the current status of the program.

The Concept Development Team must create a Concept Development Plan (CD Plan) and a Concept Document. A Concept Development Plan (CD Plan) and a Concept Document are the two main components of any CD&E project. The CD Plan is critical for the co-ordination and management of the CD&E project. It is used to assign tasks and responsibilities, monitor the progress of these tasks, integrate and synchronize efforts of multiple teams or people with different competencies and co-ordinate and integrate activities.

A universal concept structure cannot be outlined. Nevertheless, a Concept developer should consider the following generic structure for his work:

A: Introduction

- Background (to include military problem statement or capability shortfall);
- Aim (of the concept);
- Scope (of the concept; to be understood as range and limitations in time, authority, commands etc.).

B: Main Part (the conceptual solution: what needs to be done)

- Definitions;
- Requirements;
- Functions;
- Principles;
- Roles and Responsibilities;
- Assessments / Evaluation;

- Capability Elements (in terms of DOTMLPFI, as applicable);
- Resources.

Implementation (suggested concept owner, actions required to implement the concept)

Conclusion and Way Ahead (highlights benefits of the conceptual approach and outlines next steps once the concept is approved)
Some projects, mainly emergent requests, need a shorter time of development. The whole CD&E Project needs to have the flexibility to allow shortcuts, within an acceptable level of risk, as necessary to meet operational needs.

Analysis

The arsenal of the concept developer must include analysis. Operational analysis, to include static and dynamic modelling work, may provide a cost effective way to bring insights to the concept.

Analysis can determine at an early stage, stakeholders' interests in the concept and their expectations, the operational value and the feasibility of the concept, and determine possible venues for development by addressing operational validity and effectiveness through modelling. M&S takes this process a step further.

As the development of a concept progresses, the analysis activities look to accumulate evidence to determine and demonstrate the validity and increased effectiveness of the proposed solutions. The rigour of the analysis and experimentation undertaken is important to increase confidence in a conceptual solution and reduce the risk involved with its implementation.

Experimentation:

The role of experimentation is primarily to discover information, confirm or disprove a hypothesis, or provide formal validation that a concept under development will achieve its desired aim. Experimentation uses real (or nearreal) environments to provide levels of information and insight not available from other traditional means such as operational analysis or lessons learned analysis. Benefits of experimentation include:

- Reduced uncertainty (risk mitigation);
- Objective evaluation of innovations;
- Identification and possible solution of practical problems that cannot be determined through studies and analysis alone;
- Provision of empirical evidence to inform policy, budgetary and procurement decisions.

Figure 4.2 demonstrates a position of experiment in the process of conception development and transformation effects enhancement on defense capabilities. The meaning consists of strengthening creation of real capabilities and their improvement, in effectiveness of military forces operational capabilities, their building and training so thus to meet both present and future tasks as well as economic feasibility.

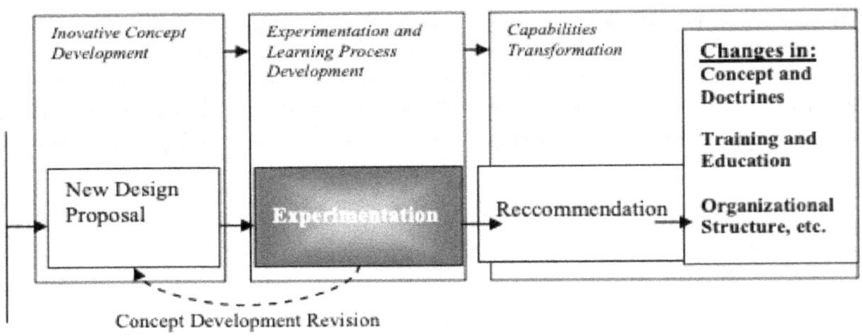

Figure 4.2 Experimentation in the process of conception development and enhancement with transformation effects on defense capabilities

Experimentation must be developed across the scale of decision making processes of the University of Defense and must be focused not only on verification and validation of new strategic intentions and conceptions but also on operational intentions and conceptions, which are essential to accomplish the OS.

Figure 4.3 shows illustratively the place of experimentation in the conception development sequence. The meaning is to improve and develop innovative operational conception and to develop an array of innovative operational capabilities to transform these conceptions into real operational capabilities.

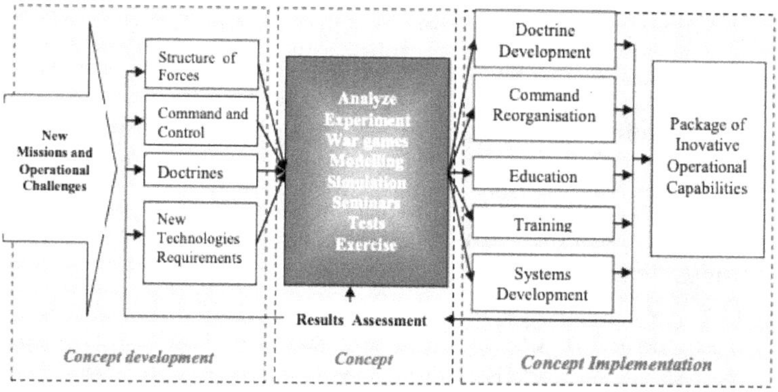

Figure 4.3 Position of experimentation in the conception development sequence.

Certain principles for an effective experimentation process must be taken. These principles are connected into three main thematic groups. The first group consists of principles needed to design a valid experiment (development of causality research capabilities, experimentation logics, requirements accomplishment to use new capabilities or change detection and the use in actual operations). The second group consists of analytic capabilities integration principles into the experimental process (the use of iterative problem formulation process, analysis and experimentation to accumulate knowledge – integration of studies, observation and experimentation, the use of varying methods to valid results confirmation). The third group is dealing with the respects needed for successful experiment (human variability, training, operational exercise, M&S, experimental management modes, collection data complex analysis, ethical, political, international and safety issues, environmental concerns, customer (creditor) communication during experimentation).

Five Components of an Experiment

All experiments—large or small, field or laboratory, military or academic, applied or pure— consist of five components [Cook and Campbell 1979]:

1. The treatment, the possible cause A, is the proposed capability, the proposed solution that is expected to influence warfighting effectiveness.

1. The possible effect B of the treatment is the result of the trial, an increase or decrease in some aspect of warfighting effectiveness.
2. The experimental unit executes the possible cause and produces an effect.
3. The trial is one observation of the experimental unit under treatment A or under the alternative ~A to the new capability to see if effect B occurred or not and includes all of the contextual conditions under which the experiment is executed.
4. The analysis phase of the experiment compares the results from one trial to a different trial.

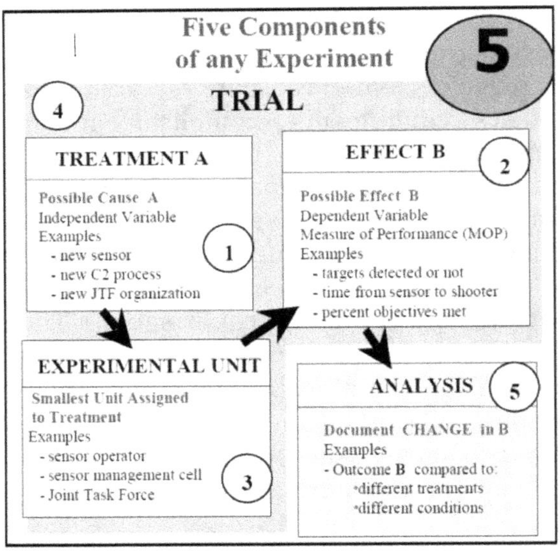

Figure 4.4 Five components of an experiment

There is a wide scale of available techniques for M&S experimentation. In experimentation the balance requirements between accuracy and validity (goal fulfilment) of any experiment must be fulfilled. Information technologies development significantly influenced modelling capabilities and contributed to more accurate results.

Most of these experiments use different kinds of simulations. Every kind has its advantages and disadvantages in regards to the four validity aspects – none of them can fulfil all four validity aspects. For this reason the variability of these four kinds of simulations are essential:

- Constructive simulations
- Analytical military games
- Human-in-the-loop simulation
- Live (field) simulation

The experimentation process consists of three main parts. The preliminary planning is followed by the experimentation programme which is followed by managing individual experiments. The preliminary planning part defines the experimentation mode to ensure meeting the goals fulfilment. This part provides a draft plan for experiment conducting and costs evaluation. The result is its specification which describes the experimentation plan in detail, defines a sequence of particular operations and identifies financial costs of individual experiments.

Conclusion

Requirements for military forces development in multinational asymmetrical operations in global framework are so urgent that it is necessary to established functional process for verifying new concepts and findings in form of experimentation to eliminate the possibility of false decisions, which can lead to extensive visible or invisible losses (material, financial, personal, but also sociological, psychological etc.), in advance.

Newly developed conception is in essence untested theory, can offer whole new approaches, processes, requirements etc. However, with this there are no experiences. This is the reason to verify these approaches and processes before their implementation into practice. One of the methods used to verify plausibility of the new conceptions in military forces environment is experimental method. The primary role of experimentation is to determine whether conception in development can accomplish set goals or not. The results then inform the conception author whether whole conception or its parts can achieve desired utility or not and it helps identify and solve problems that cannot be dealt with other means e.g. studies or analysis.

4.2 Radical Innovation tools

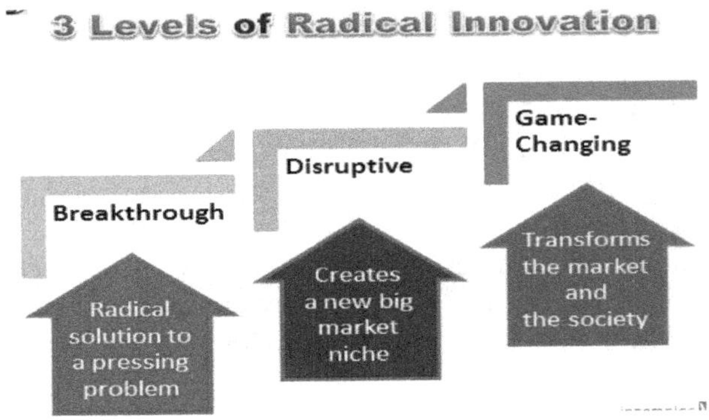

Figure 4.5 3 Levels of radical innovation

Radical innovation, concerned with exploration of inventions and new technology, is fundamentally different from incremental innovation that is concerned with exploitation of existing technology.

The **3 levels** of radical innovation:

1. Breakthrough innovation delivers a creative radical solution to a challenging problem.
2. Disruptive innovation creates a new big market niche.
3. Game-changing innovation transforms the market and the society.

HBS Definition

According to Harvard Business School, "Radical innovation is a product, process, or service with either unprecedented performance features or familiar features that offer potential for → **significant improvements** in performance and cost."

Distinctive Features of **Radical Innovation Projects**

Produce an entirely new set of performance features
and/or
Produce improvements in known performance features of five times or greater
and/or
Produce a significant (30% or greater) reduction in cost

4.3 Rapid Prototyping, Prototyping Techniques

Rapid prototype or **rapid prototyping** is a relatively new term and in its simplest form, the process of creating prototypes quickly to visually and functionally evaluate an engineering product design.

In an **engineering product design** context, a prototype is a preliminary version of the end-product and used to evaluate the design, test the technology or analyse the working principle which in turn provides product specification for a real working system.

Prototypes are an integral part of engineering product design and more importantly in an overall new product development process. Rapid prototyping can be used at any stage of the product development cycle or for any component or sub-component and can be repeated numerous times along the new product design process.

NPD Stage Types

In modern-day engineering product design, prototyping process of build, review and refine, fits into all four major stages of the design process (product planning, conceptual design, embodiment design and detailed design).

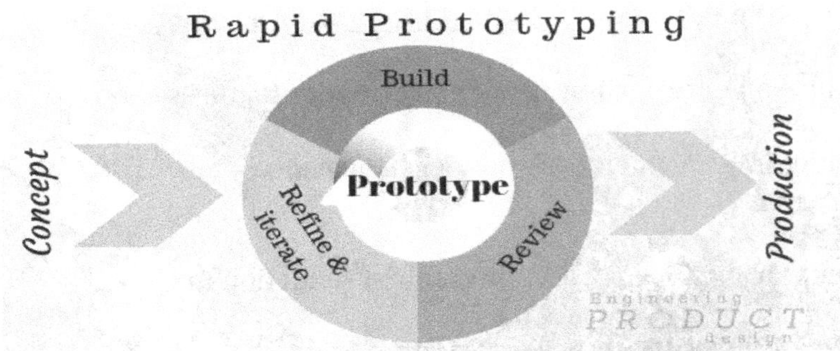

Figure 4.6 Rapid prototying

- Proof of concept prototypes
- Demonstration or presentation model prototypes
- Functional prototypes
- Aesthetic or industrial design prototypes
- Final factory sample
- Alpha & beta build prototypes

Types of Rapid Prototyping Techniques

Choosing the right rapid prototyping technology is critical to the success of a prototype. Each rapid prototyping technique has its own compromise in terms of cost, speed, material compatibility of the feature, fidelity level and development stage.

Rapid prototyping doesn't need to be limited to one process, one can use more than one manufacturing techniques to assemble a prototype.

Following are the types of rapid prototyping technology available for engineering product designers:

- Additive manufacturing
- Stereolithography (SLA)
- Selective laser sintering (SLS)
- Direct metal laser sintering (DMLS)
- Fused Deposition Modelling (FDM)
- Binder jetting

- Poly jetting
- Other techniques
    - CNC Machining Prototyping
    - Vacuum casting
    - Investment casting

Advantages and Disadvantages of Rapid Prototyping

Like any manufacturing process or design stage, prototyping and rapid prototyping have their own pros and cons.

Advantages of Rapid Prototyping

- Reduced design & development time
- Reduced overall product development cost
- Elimination or reduction of risk
- Allows functionality testing
- Improved and increased user involvement
- Ability to evaluate human factors and ergonomics

Disadvantages of Rapid Prototyping

- Lack of accuracy
- Added initial costs
- Some rapid prototyping processes are still expensive and not economical
- Material properties like surface finish and strength cannot be matched
- Requires skilled labour
- The range of materials that can be used is limited
- Overlooking some key features because it cannot be prototyped
- End user confusion, customers mistaking it for the finished project/developer misunderstanding of user objectives

4.4 User Testing and Feedback

User testing is a technique employed in design to get a website, product/MVP (Minimum Viable Product), feature or a prototype assessed through the real users.

User testing is crucial as it lets the design team find any discord in the user experience that they are designing so that any issues can be addressed and rectified before the final product goes live. Identifying and fixing issues at the earlier stages indeed cuts down the long-term cost.

Usability tests provide feedback on how real customers use particular parts of your product to complete specific tasks. By watching them perform the tasks, you can learn whether your product is well suited to the task, and how it might be improved.

4.5 Feedback Analysis

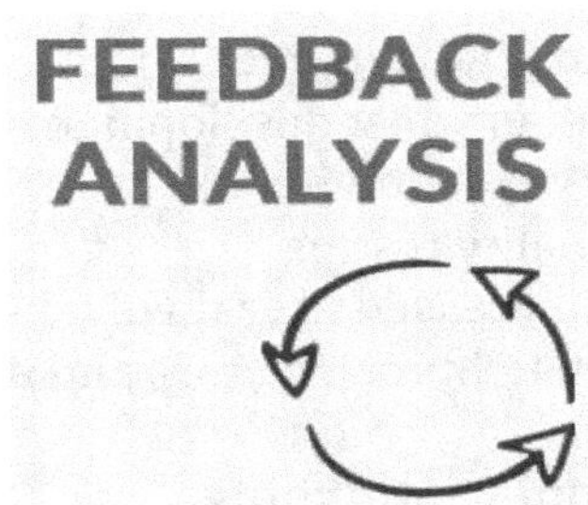

Figure 4.7 Feedback analysis

Feedback analysis is the process of setting up objectives for yourself and reviewing how you performed.

It is the best way to get out of your comfort zone, but still being in control. It's also a great way to set small personal growth goals and reach them.

Five Steps to Perform Effective Feedback Analysis

1. Categorize Feedback
2. Find Root Causes
3. Plan Further Actions
4. Inform Teams
5. Use Automated Tools

4.6 Iterating and Refining

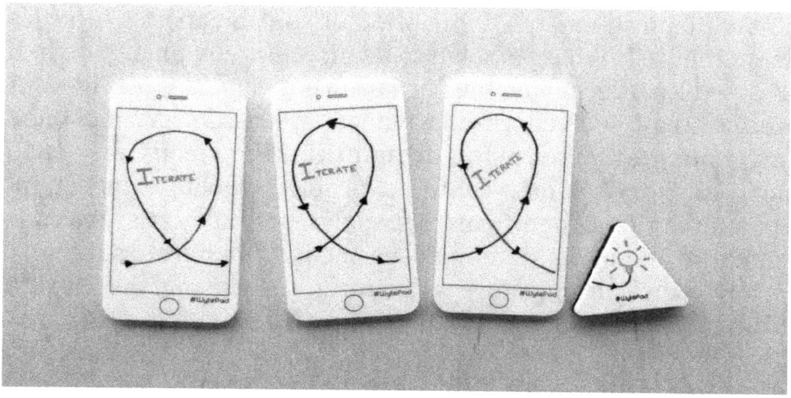

Figure 4.8 Iterating and Refining

The Iterative Design Approach

Creating design in an iterative design approach means refining the design through a number of iterations. Each iteration can consist of several activities, e.g. designing, prototyping and testing. First you create design suggestions and pick one or more to prototype. The prototypes are then used to test the design, and through tests you will gather feedback on the design. This feedback is then used for the next iteration, refining the design, improving the prototype and testing this again. This loop can be repeated as many times as necessary until the design solves whatever requirements you have set for it and the users' needs are solved appropriately.

The benefits of taking an iterative approach to design are many. It encourages experimentation and trying out design alternatives early on in a project, allowing for the design team to quickly identify risks and concerns that needs iterations to end up with a suitable design. It also encourages more involvement of users to make informed decisions in early iterations. This can save time and resources in the long run as features, that are tested to be unnecessary, can be avoided, and features that are shown to be essential can be implemented instead.

4.7 Business plan

A business plan is a written description of your business's future, a document that tells what you plan to do and how you plan to do it. If you jot down a paragraph on the back of an envelope describing your business strategy, you've written a plan, or at least the germ of one. Creating a **business plan using design thinking** means approaching your plan from a human-centered, problem-solving mindset. Instead of jumping straight into market analysis or financials, you start with empathy and understanding user needs, then iterate solutions. Here's how you can structure a business plan using the **Design Thinking process**:

1. Empathize: Understand Your Users

Goal: Gain deep insight into your customers' problems, needs, and emotions.

- **User Research**: Interviews, surveys, observations.
- **User Personas**: Create 2–3 fictional yet realistic personas that represent your target users.
- **Pain Points**: What problems are they facing? What frustrates them?

Example:
"Small café owners struggle to manage online orders while juggling in-store customers.

2. Define: Clearly Articulate the Problem

Goal: Synthesize your research into a clear and actionable problem statement.

- **Problem Statement Template**:
  "[User] needs a way to [user's need] because [insight]."
- Define your value proposition: What job are you helping users get done?

Example:
"Café owners need an integrated order management system because juggling digital and in-person orders slows down service and frustrates customers."

3. Ideate: Generate Solutions

Goal: Brainstorm as many solutions as possible, without judgment.

- Use methods like mind mapping, "how might we" questions, or SCAMPER.
- Narrow down to 1–2 high-potential ideas.

Example:
Idea: A smart tablet-based dashboard that integrates online and in-person orders in one view, with voice prompts and status updates.

4. Prototype: Build a Testable Version

Goal: Create a lightweight version of your product or service.

- Sketches, mockups, landing pages, or clickable prototypes.
- Focus on core functionality—not polish.

Example:
A clickable Figma prototype of the dashboard, or a demo video simulating the process.

5. Test: Get Real Feedback

Goal: Validate your assumptions and improve your solution based on feedback.

- Conduct user testing sessions.
- What worked? What confused users? What should be improved?
- Iterate.

Business Plan Sections (Design Thinking-Aligned)

Here's how a **formal business plan** maps onto this design thinking approach:

Business Plan Section	Design Thinking Phase	Description
Executive Summary	All phases (summary)	Brief overview of the problem, solution, market, and financials.
Problem Statement	Empathize + Define	Deep insights into the user and their core pain points.
Solution	Ideate + Prototype	Describe your innovative, user-centered product or service.
Market Analysis	Empathize	Define target market, trends, and customer segmentation.
Business Model	Ideate	How will you deliver value and make money?
Go-to-Market Strategy	Prototype + Test	Launch plan, marketing, sales, and user feedback loops.
Operations & Tech	Prototype	How the product will be built and delivered.
Financial Plan	Test (Assumptions)	Forecast revenue, costs, and break-even analysis.
Milestones & KPIs	Test	Metrics to track user adoption and business growth.

4.8 Human Desirability- Technical Feasibility - Business Viability, Brand Identity, Mission and Vision Statements

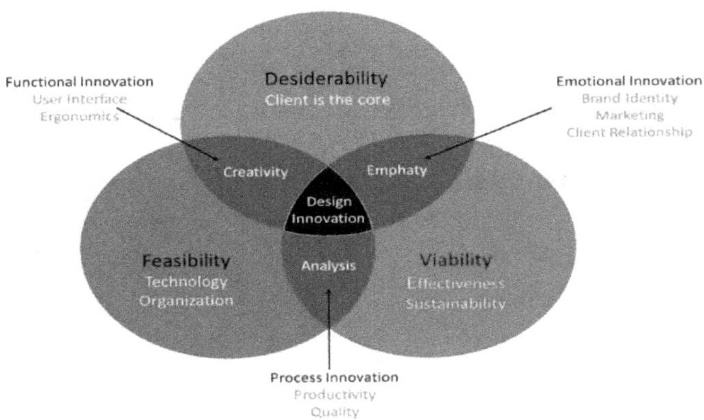

Figure 4.9 Human desirability

The goal of Design Thinking is enabling the identification of an innovative solution to a problem/opportunity which satisfies 3 core criteria: desirability (by the market or the actors), feasibility (technology or organization), and viability:

- **Desirability**: empathize with clients, turning their needs and desires into your own. The aim of this phase is making clients see what they could need and not trying to convince them "the product/service is useful". Look and ask questions about their behaviour (empathy mapping).

- **Feasibility**: identify themes and models, through observation and field activities, trying to build relationships and deepening insights. Once the process is finished, you explore the product/service ideas and test them through iterative prototyping and quick experimentation to check how consumer react, adjusting the product, its price, or positioning.

- **Viability**: define the winning solution, in terms of sustainability and profitability, identifying any activity or resource the company will actually have to produce to communicate, distribute and deliver this solution.

Design Thinking inside a company enables the discovery of new market opportunities and to guide clients' problems and desires by creating innovative and effective offers. Thanks to its processes and tools, it enables you to optimize your **available resources**, either time, money, or people involved. It also fosters a positive and proactive environment, and develops a sense of belonging inside the company.

Design Thinking helps in developing **decision making processes**, as well as **the quality of the made decisions**. It is a fundamental help for decision makers when they have to make vital decisions for their company, while reducing risks.

Nowadays, the designing process is still trapped inside the "little creative room" in many companies, a room where designers are not actively and effectively involved in developing the company business, since the project is seen more as a **tool** than as a **strategy**. *Design-oriented* companies, like Apple, Microsoft, IKEA, Alessi, and Olivetti (whose founder had seen this truth many years before anyone else) have been able to integrate the design function at the executive body level, to make strategic decision relevant for the future of the company.

It is fundamental, in English and American companies, to understand and share **how** you get to

- the conception of a project
- well balanced solutions and important strategic decisions
- the definition of a solid foundation which allows you to invest money

Brand identity is the visible elements of a **brand**, such as colour, design, and logo that identify and distinguish the **brand** in consumers' minds. **Brand identity** is distinct from **brand** image.

Mission: We ignite opportunity by setting the world in motion.

Vision: Smarter transportation with fewer cars and greater access. Transportation that's safer, cheaper, and more reliable; transportation that creates more job opportunities and higher incomes for drivers.

Module 5

Communicating and Presenting

5.1 Communicating and Presenting

A presentation is a means of communication that can be adapted to various speaking situations, such as talking to a group, addressing a meeting or briefing a team.

A presentation can also be used as a broad term that encompasses other 'speaking engagements' such as making a speech at a wedding, or getting a point across in a video conference.

To be effective, step-by-step preparation and the method and means of presenting the information should be carefully considered.

A presentation requires you to get a message across to the listeners and will often contain a 'persuasive' element. It may, for example, be a talk about the positive work of your organisation, what you could offer an employer, or why you should receive additional funding for a project.

The Key Elements of a Presentation

Making a presentation is a way of communicating your thoughts and ideas to an audience and many of our articles on communication are also relevant here.

What is Communication?

Consider the following key components of a presentation:

Ask yourself the following questions to develop a full understanding of the context of the presentation.

When and where will you deliver your presentation?
There is a world of difference between a small room with natural light and an informal setting, and a huge lecture room, lit with stage

lights. The two require quite different presentations, and different techniques.

Will it be in a setting you are familiar with, or somewhere new?

If somewhere new, it would be worth trying to visit it in advance, or at least arriving early, to familiarise yourself with the room.

Will the presentation be within a formal or less formal setting?

A work setting will, more or less by definition, be more formal, but there are also various degrees of formality within that.

Will the presentation be to a small group or a large crowd?

Are you already familiar with the audience?

With a new audience, you will have to **build rapport** quickly and effectively, to get them on your side.

What equipment and technology will be available to you, and what will you be expected to use?

In particular, you will need to ask about microphones and whether you will be expected to stand in one place, or move around.

What is the audience expecting to learn from you and your presentation?

Check how you will be 'billed' to give you clues as to what information needs to be included in your presentation.

Presenter

The role of the presenter is to communicate with the audience and control the presentation.

Remember, though, that this may also include handing over the control to your audience, especially if you want some kind of interaction.

Audience

The audience receives the presenter's message(s). However, this reception will be filtered through and affected by such things as the listener's own experience, knowledge and personal sense of values.

The message or messages are delivered by the presenter to the audience.

The message is delivered not just by the spoken word (**verbal communication**) but can be augmented by techniques such as voice projection, body language, gestures, eye contact (**non-verbal communication**), and visual aids.

The message will also be affected by the audience's expectations. For example, if you have been billed as speaking on one particular topic, and you choose to speak on another, the audience is unlikely to take your message on board even if you present very well. They will judge your presentation a failure, because you have not met their expectations.

Reaction

The audience's reaction and therefore the success of the presentation will largely depend upon whether you, as presenter, effectively communicated your message, and whether it met their expectations.

As a presenter, you don't control the audience's expectations. What you can do is find out what they have been told about you by the conference Organisers, and what they are expecting to hear. Only if you know that can you be confident of delivering something that will meet expectations.

Method

How will the presentation be delivered?

Presentations are usually delivered direct to an audience. However, there may be occasions where they are delivered from a distance over the Internet using video conferencing systems, such as Skype.

It is also important to remember that if your talk is recorded and posted on the internet, then people may be able to access it for several years. This will mean that your contemporaneous references should be kept to a minimum.

Impediments

Many factors can influence the effectiveness of how your message is communicated to the audience.

For example, background noise or other distractions, an overly warm or cool room, or the time of day and state of audience alertness can all influence your audience's level of concentration.

As presenter, you have to be prepared to cope with any such problems and try to keep your audience focused on your message.

1. Communicating Research Insights

What is it?

Translating raw user data (from interviews, observations, etc.) into clear, meaningful insights.

Why does it matter?

Empathy is the foundation of design thinking. Clearly communicating user needs and pain points ensures the team stays user focused.

It includes:

- Synthesizing data into key themes
- Creating user personas, empathy maps, journey maps
- Presenting insights in a way that inspires action

2. Storytelling in Design

What is it?

Using narrative techniques to explain design challenges, user needs, and the value of solutions.

Why does it matter?

A compelling story helps others connect emotionally to the user and see the value of your idea.

It includes:

- Framing the problem as a story
- Sharing user journeys
- Using storytelling structures (e.g., problem-solution-impact)

3. Visual Communication

What is it?

Using sketches, diagrams, wireframes, and prototypes to make ideas visible and easier to understand.

Why does it matter?

Visuals help align teams, reduce ambiguity, and make abstract ideas concrete.

It includes:

- Sketching and diagramming ideas
- Using design tools (e.g., Figma, Miro, Canva)
- Creating low- and high-fidelity prototypes

4. Presenting Ideas to Stakeholders

What is it?

Crafting clear, persuasive presentations of your ideas for feedback, support, or decision-making.

Why does it matter?

Great ideas can fail without effective communication. You must convince stakeholders of their value.

It includes:

- Structuring your presentation (problem, solution, value)
- Using storytelling and visuals
- Tailoring messages to different audiences (users, executives, designers)

5. Prototypes as Communication Tools

What is it?

Using tangible representations of your idea to gather feedback or explain a concept.

Why does it matter?

Prototypes allow others to experience your idea, often revealing insights and sparking collaboration.

It includes:

- Creating mockups, paper models, or interactive demos
- Using prototypes in user testing and presentations
- Explaining how the prototype reflects user needs

6. Pitching Design Solutions

What is it?

Delivering a focused, persuasive summary of your solution — usually at the end of the design thinking process.

Why does it matter?

A strong pitch can win buy-in, funding, or implementation support.

It includes:

- Crafting a concise narrative
- Using data and visuals to support your claims
- Managing time and answering questions effectively

7. Collaborative Communication

What is it?

Working and presenting with others as a team, especially during ideation and solution phases.

Why does it matter?

Design thinking is a team sport. Clear and open communication ensures alignment and builds trust.

It includes:

- Co-presenting ideas
- Facilitating feedback sessions or workshops
- Managing team dynamics and roles in presentations

5.2 Marketing Pitch

In **Design Thinking**, a **marketing pitch** is a short, persuasive presentation that communicates your **user-centered solution** to a target audience — such as stakeholders, investors, or customers. It's the final step where you showcase your idea, explain how it solves a real problem, and invite others to support, adopt, or invest in it.

- It's not just about selling — it's about storytelling, empathy, and **showing the impact** of your design solution.
- Why It Matters in Design Thinking
- Design thinking revolves around understanding user needs, creating human-centered solutions, and iterating through feedback. A good pitch helps you:
- **Tell the story of the user**
- **Demonstrate the solution's value**
- **Inspire action** — support, funding, testing, or implementation
- Your pitch is proof that your solution is **grounded in empathy**, backed by research, and ready for the real world.

Elements of a Marketing Pitch in Design Thinking

- **Start with Empathy (The Hook)**
- Begin with a user story or insight.
- Make the audience care about the problem.
- **Define the Problem**
- Clearly articulate the need or pain point discovered through research.
- **Present Your Solution**
- Show how your idea directly addresses the problem.
- Use prototypes, visuals, or demos.
- **Highlight the Value Proposition**
- Explain what makes your solution unique, effective, or innovative.
- **Market Opportunity**
- Briefly present who the users are and the potential for scale.
- **Call to Action**
- Ask for what you need: investment, feedback, adoption, support.

Example: Design Thinking Pitch

"Meet Amina, a college student who spends 3 hours daily looking for quiet study spaces. After interviewing 20 students, we found this is a common frustration. So, we designed **StudyFindr**, an app that helps students instantly locate nearby open study spaces in real-time. Our prototype reduced search time by 75%. With 1 million students in the region, there's a huge opportunity to improve productivity campus-wide. We're asking for support to launch a beta version at three universities."

Simply defined, it's a line of talk that is designed to persuade someone, such as a presentation to sell a product or service. But it can also be about selling yourself, to gain new clients or to promote your business.

1. Start with your personal marketing pitch
2. Customize your personal pitch as necessary
3. Build a marketing pitch for your business
4. Customize the business pitch
5. Consider what's best for your client

6. Learn from others
7. Customize for the medium
8. Gather proof of your value
9. Don't overdo it
10. Check and double-check
11. Don't forget to follow up

5.3 Process Documentation

Process documentation is a **detailed description** of how to execute a process. The process document outlines the exact steps needed to complete a task or process from start to finish.

The area of process documentation triggers on how employee members perform the process, and not what the process is. Process documentation is important for any business because it enhances consistency and lets your staff learn from both their **successes and their mistakes**.

Benefits of Process Documentation

Process documentation helps in:

- Saving **time & money**
- Reducing errors
- Creating standardization
- Improving employee **onboarding**
- Creating workplace efficiencies
- Finding additional ways to further improve processes
- Improving workflows **analytically**

Process documentation is an **ongoing** documentation of a process while a project or task is being carried out and in no way, is the final report. It helps analyze every step of a process and provide context to improve the process.

Why Do You Need Process Documentation?

When it comes down to creating new processes or refining existing ones, there's no better way to do it than to document the processes.

The following are **primary benefits** that an organization seeks to achieve by documenting their processes:

1. Improving a Process
2. Use it as Training Material
3. Reduce Operational Ambiguity
4. Preserve Key Knowledge
5. Analysis

How to Create Process Documentation?

Process documentation's complexity depends upon the **scale of the process** being documented. If it's a shorter process, it's relatively easier to document a process as it involves a smaller number of people, stakeholders and the overall number of steps.

However, if it's a broad-scope process, involving internal as well as external stakeholders from different departments and specialties, process documentation can turn out to be **fairly complex**.

Therefore, we need to have certain guidelines or roadmap in place to make sure that process documentation is carried out effectively, irrespective of process size and scope.

The following are some **key steps** your team should incorporate while preparing for a business process documentation.

1. Scope

First and foremost, determine the scope of the process document. The scope answers questions like which process will be covered, what are the goals of the project, what changes are to be made (if any), etc.

2. Resources needed

Describe the resources that the team needs in order to complete the process you want to describe (inputs required). You also need to describe the outputs or what is **being produced** at the end of process documentation.

3. Know your audience

If you don't know who your audience is and what their expectations are from the process documentation, you are wasting your time. Only when you are **familiar** with the audience can you create a document that meets their requirements.

4. Describe the team involved

Describe the team members involved in a process. Describe their roles and responsibility to establish clarity.

5. Collect information

Information is often collected through a **brainstorming session** about the steps required to complete an activity amongst the team members involved. It can also involve the views or expertise of stakeholders and outsiders.

6. Putting it all together

After you have made sure that all the steps required to perform a process from start to finish are noted, it's time to organize them in a sequential list format.

Make sure the steps are organized in the **actual manner** in which they work (process) is done. Write down the steps in a clear, concise way to be useful to your co-workers. Make the process documents easy to read by using bullets, headings, tables, charts, etc.

7. Flowchart

It's now time to visualize the list of activities for your team in terms of a flowchart. Visually representing the steps involved in a process provides an easy way for outsiders and people from other departments to understand how a process is carried out. It also aids new employees to soak in information quickly and get up to speed with the work.

8. Get feedback

Share the process documentation with fellow colleagues and stakeholders and have them try to perform the process on their own.
This is a critical step in your documentation process as it helps to determine **potential loopholes** or corrections in the documentation by allowing outsider's perspective to the document.
The visual representation of the process helps team members to see the process in a different light and find out if they missed something.

9. Improve

It's time to optimize the steps and add any **improvements** suggested by colleagues, stakeholders or fellow team members. Rinse and repeat until your process is completely optimized and efficient. Keep adding changes to the document when you find a new strategy technique or tactic to improve your process.

10. Make it available

Once you are satisfied with the results, make the documentation available to relevant employees and stakeholders of the company. Ask them to freely provide their valuable suggestions and feedback.

5.4 Photo Documentation

Photographs provide a qualitative, and potentially semi-quantitative, record of conditions in a watershed or on a water body. Photographs can be used to document general conditions on a reach of a stream during a stream walk, pollution events or other impacts, assess resource conditions over time, or can be used to document temporal progress for restoration efforts or other projects designed to benefit water quality. Photographic technology is available to anyone and it does not require a large degree of training or expensive equipment. Photos can be used in reports, presentations, or uploaded onto a computer website or GIS program. This approach is useful in providing a visual portrait of water resources to those who may never have the opportunity to actually visit a monitoring site.

Recording Information

Use a systematic method of recording information about each project, photo point, and photo. The following information should be entered on the photo-log forms or in a dedicated notebook: • Project or group name, and contract number (if applicable, e.g., for funded restoration projects)

- General location (stream, beach, city, etc.), and short narrative description of project's habitat type, goals, etc.
- Photographer and other team members
- Photo number
- Date
- Time (for each photograph)
- Photo point information, including:
    - Name or other unique identifier (abbreviated name and/or ID number)

    - Narrative description of location including proximity to and direction from notable landscape features like roads, fence lines, creeks, rock outcrops, large trees, buildings, previous photo points, etc. - sufficient for future photographers who have never visited the project to locate the photo point
    - Latitude, longitude, and altitude from map or GPS unit
- Magnetic compass bearing from the photo point to the subject
- Specific information about the subject of the photo
- Optional additional information: a true compass bearing (corrected for declination) from photo point to subject, time of sunrise and sunset (check newspaper or almanac), and cloud cover. When monitoring the implementation of restoration, or Best Management Practices (BMP) projects, include or attach to the photo-log a narrative description of observable progress in achieving the goals of the project. Provide supplementary information along with the photo, such as noticeable changes in habitat, wildlife, and water quality and quantity. Archive all photos, along with the associated photo-log information, in a protected environment.

5.5 Feedback and Critique

Criticism focuses on what we don't want. It focuses on what has happened in the past, and can sometimes make you feel helpless since you can't turn back time to change things. Criticism points out weaknesses and can make you feel deflated. The subtle message is, "You are the problem." Researchers John and Julie Gottman at the University of Washington have discovered that criticism is actually one of the worst things you can ever do for a relationship. They call it a Horseman of the Apocalypse.

In contrast, feedback is one of the best things we can do for our relationships. Good feedback focuses on what we do want: the behaviors, the characteristics and emotions, the choices, the teams that we most want as part of our workplace. It may be given out of a concern about a problem that has already happened, but it is framed in a way that focuses on the future. So, you have the opportunity to improve and show growth.

Feedback intentionally builds up strengths, in fact, you often make a point of sharing what a person has done right, or what strengths they're working with, as you help them focus on what is wanted for the future. As you can imagine, this kind of conversation can be inspiring. The message is, "We can make this better, together."

5.6 Logo and Graphics Generation

A logo is more than a static symbol. It carries the significance of a brand's visual identity and without one; a brand would lose its meaning. To leave an everlasting impression on a consumer, having an exceptional logo is crucial. A logo has the power to represent all the values a brand upholds.

But before we get the considerations of designing a logo, let's understand why logos are so important.

Why are Logos so Important?

Anyone starting a business would have gotten the suggestion to invest in getting a logo although some people fail to realize that having a logo is just the initial step of building a memorable brand identity. A logo is the keystone of your brand and no matter what type of brand or business you own, it is important to have a professionally designed logo. Without one, your audience will not be able to identify your company, let alone allow you to outweigh your competitors.

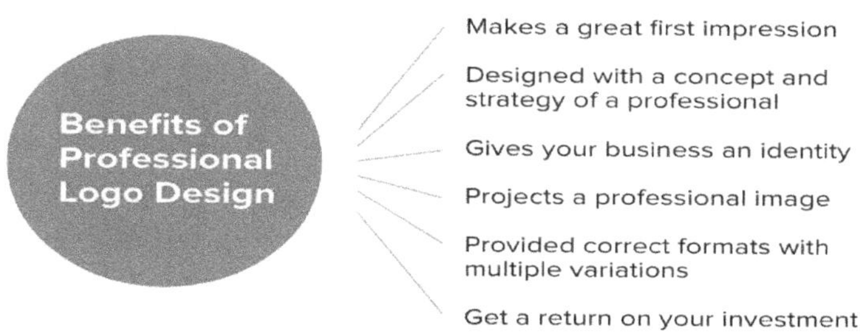

a) A logo is the face (identifier) of your business

We tend to picture a specific brand with its logo, and this is related to your experience and connection with the company. Whenever you see a logo of a brand you have previously associated with, you cannot help but recall the interactions you have had with that certain brand.

b) Logos drive consumer decision-making

We are able to make a judgment about the brand's products and services, from just a single glance at the logo. This is connected to the prior experience we have had with the brand, as well as the visual itself. If the logo matches your perception about the type of company you think it is, it can influence the decisions you make and the expectations you have from them.

c) Logos allow for quick brand recognition

Words are not enough to convey what a brand has to offer, and words are difficult to memorize. As the age old saying goes, a picture is worth 1000 words. A brand's visual identity needs to be unique so the brand can be quickly & easily identified.

d) Logos make you stand out from your competition

A logo is directly associated with a business and what its services or products are. Countless companies could be offering the same services as yours but what sets you apart from the rest is solely your logo & brand. This provides you with the option to come up with a design that leaves a lasting first impression.

The Graphic Generation Tool is used to create, open, save, customize, and configure Microsoft® Silverlight® based graphics. The Graphic Generation Tool provides a palette of graphic elements, known as the Graphic Enterprise Library (GEL) that lets you create detailed graphic representations of your building automation system. The tool also lets you communicate with Metasys hosts and child devices. In turn, this communication allows you to be more productive when creating, saving and modifying graphics because the graphics communicate directly with their Metasys system.

CASE STUDY 1: RURAL DEVELOPMENT

Development may mean different things to different people. It implies a change that is desirable. What is desirable at a particular time, place and in a particular culture may not be desirable at other places, or other times or culture Hence, there is no universally accepted definition.

However, it refers to a better quality of life encompassing at least three components:

- Self-sustenance
- self – esteem and freedom

Development implies a process of change towards achieving a better quality of life, encompassing at least three components: Self-sustenance, the basic necessities of life including food, clothes, shelter, primary health care and security of life and property. Self – esteem, self-respect, dignity, or honour for every person and every nation. Freedom- political, economic and cultural

As long as a society is bound by the servitude of men to nature, ignorance, other men, institutions, and dogmatic beliefs, it cannot claim to have achieved the goal of 'development'

IMPORTANCE

- Rural development is the backbone for any country's economic development, and it helps the economy to grown and sustain.
- 80-95% people in Rural do agriculture: and it is a popular belief that economic development taken place because of rapid industrialization.
- But the industrial development itself can't take place without agriculture sector.
- This sector of development of economy is important to feed the nation.
- Rural development is important not only for the majority of the population residing in a rural area but the growth of rural activities is necessary to stimulate the speed of overall economic expansion of the nation.

CHALLENGES

- High incidence in rural areas of income poverty, is not the only dimension of poverty.
- Most development indicators related to education, health and gender equality show the same pattern.
- So, while urban poverty is growing fast and matters enormously, **fighting poverty today means first a transformation of rural lives and livelihoods**.
- Such a strategy would also have a positive impact on urban poverty and on forced out-migration, which are often direct consequences of lack of rural opportunities.
- Despite the diversity and complexity of the livelihoods of the rural poor, - the majority of rural households depend mostly on agriculture and agriculture related small industries and services.

5-Y ANALYSIS and GOALS

Goal 1 Eradicate extreme poverty and hunger

Target 1: Halve, between 2020 and 2025, the proportion of people whose income is less than one dollar a day.
Target 2: Halve, between 2020 and 2025, the proportion of people who suffer from hunger.

Goal 2 Achieve universal primary education

Target 3: Ensure that, by 2025, children everywhere, boys and girls alike, will be able to complete a full course of primary schooling.

Goal 3 Promote gender equality and empower women

Target 4: Eliminate gender disparity in primary and secondary education, preferably by 2025, and in all levels of education no later than 2030.

Goal 4 Reduce child mortality

Target 5: Reduce by two thirds, between 2020 and 2022, the under-five mortality rate.

Goal 5 Develop agricultural activities

Target 6: Introduce proper irrigation techniques

SOLUTIONS

The need for reducing poverty and to develop the standard of living in rural areas around the world is enormous. According to our analysis, the first four goals we have, is possible only with the change in attitude of the society and the mind-set of people. Whereas the fifth goal can be implemented with the basic concept of innovation. Ideas for new approaches have to be created, developed and implemented. Design thinking and methods combined with innovation in practice and management – as Design Inspired Innovation – could be such a concept to provide for rural people to empower themselves and improve their living conditions.

OUR FINAL SOLUTIONS

Solution 1: Freshwater is needed for crop and energy production, industrial fabrication as well as human and ecosystem needs. 69% of the total extracted freshwater is used by agriculture sector, whereas 19% is used by industrial sector and the rest in used by domestic segment. Therefore, water can be considered as a critical need in agriculture sector for future global food security However, continued increase in demand for water by domestic and industrial sectors and greater concerns for environmental quality have create a challenge to every country to reduce the farm water consumption and sustain the fresh food requirement. Consequently, there is an urgent need to create strategies based on science and technology for sustainable use of water. Industrialist and researchers are working to build efficient and economic automatic systems to control water usage in order to reduce much of the wastage.

Irrigation is an artificial application of watering the land for agricultural production. The requirement of water to the soil depends on soil properties such as soil moisture and soil temperature. Effective irrigation can influence the entire growth process and automation in irrigation system using modern technology can be used to provide better irrigation management. In general, most of the irrigation systems are manually operated. These traditional techniques can be replaced with automated techniques of irrigation in order to use the water efficiently and effectively.

A sensor based automated irrigation system provides promising solution to farmers where the presence of a farmer in field is not compulsory during irrigation process. Arduino is a flexible programmable hardware platform and designed to control the circuit logically. So, the solution that can handle the water flow. The solution is **Automatic Irrigation System**

"AUTOMATIC IRRIGATION SYSTEM"

MATERIALS REQUIRED

1. Arduino UNO
2. LCD Display
3. Soil Moisture Sensor
4. Temperature sensor
5. Water level switch
6. 5V Relay
7. Water pump
8. Potentiometer

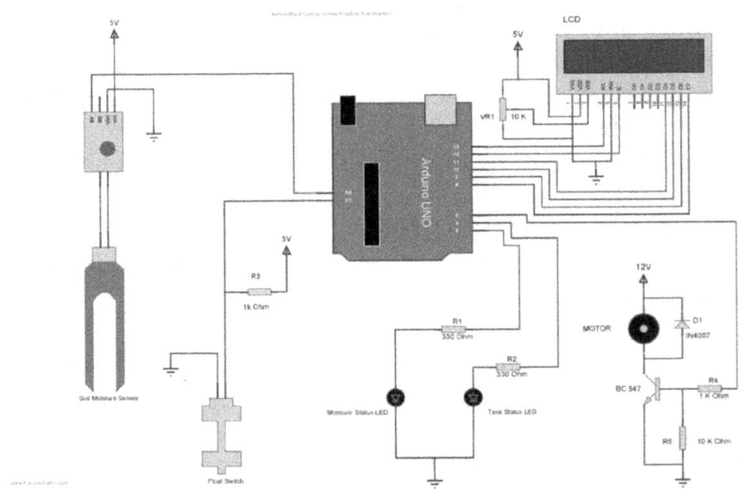

DESIGN PHASE AND WORKING METHOD

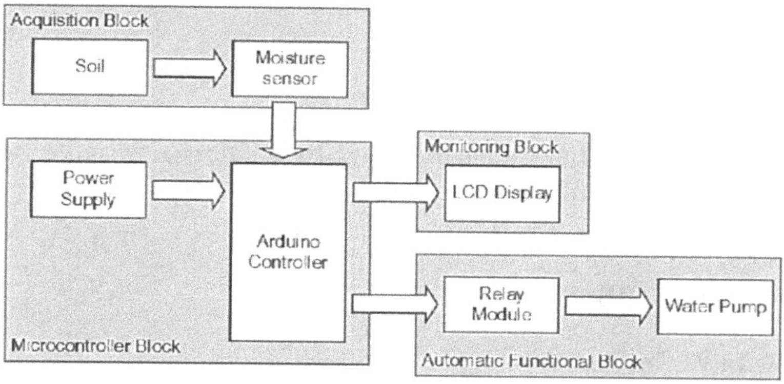

CODE

```
#include <Servo.h>
Servo myservo;int m=0;
int n=0;
intpos = 0;
void setup() {
// put your setup code here, to run once:
pinMode(A0, INPUT_PULLUP);
pinMode(A1, INPUT_PULLUP);
pinMode(8,OUTPUT);
Serial.begin(9600);
myservo.attach(9);
}
void loop() {
// put your main code here, to run repeatedly:
int m= analogRead(A0);
int n= analogRead(A1);
Serial.println(m);
delay(10);
Serial.println(n);
delay(200);
if (m>=980)
{
myservo.write(90); // tell servo to go to position in variable 'pos'
digitalWrite(8, HIGH);
delay(1000);
}
```

```
else if(m<=970)
{
digitalWrite(8, LOW);
}
if (n>=980)
{
myservo.write(0); // tell servo to go to position in variable 'pos'
digitalWrite(8, HIGH);
delay(1000);
}
else if(n<=970)
{
digitalWrite(8, LOW);
}
else
{
digitalWrite(8, LOW);
}
```

SOFTWARE USED

Arduino IDE

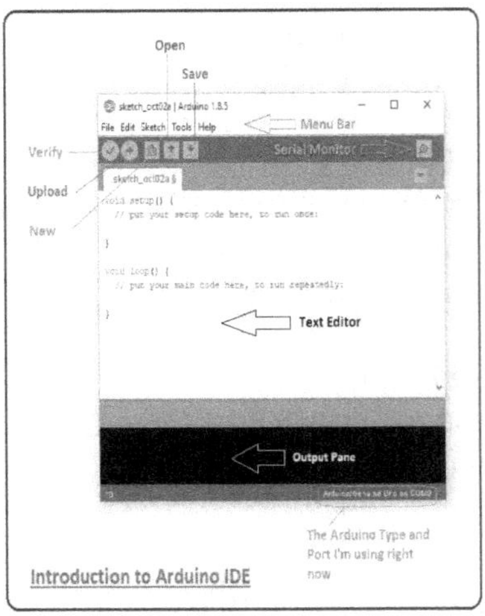

PROTOTYPE

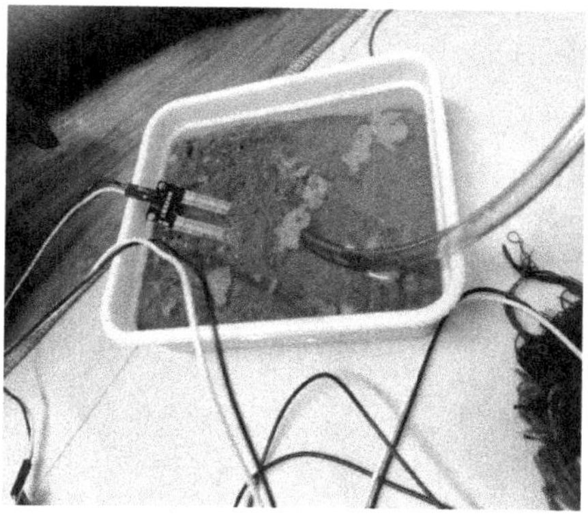 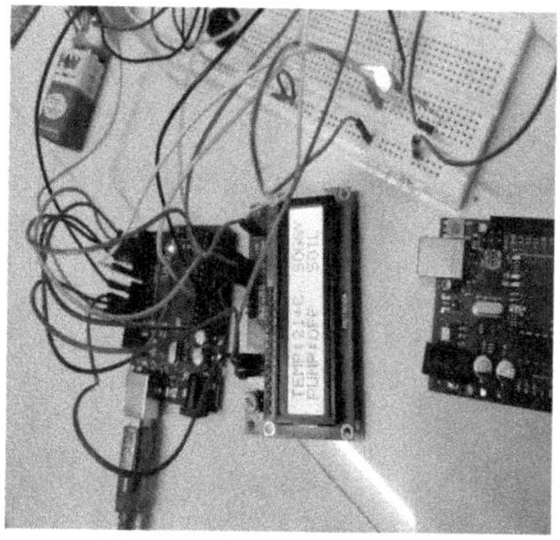

Solution 2: Internet of Things (IoT) technology has brought revolution to each and every field of common man's life by making everything smart and intelligent. IoT refers to a network of things which make a self-configuring network. The development of Intelligent Smart Farming IoT based devices is day by day turning the face of agriculture production by not only enhancing it but also making it cost-effective and reducing wastage. The aim / objective of this project is to propose a Novel Smart IoT based Agriculture Stick assisting farmers in getting Live Data (Temperature, Soil Moisture) for efficient environment monitoring which will enable them to do smart farming and increase their overall yield and quality of products. **The agriculture stick** being proposed via this paper is integrated with Arduino Technology, Breadboard mixed with various sensors and live data feed can be obtained online from Thingsspeak.com. The product being proposed is tested on Live Agriculture Fields giving high accuracy over 98% in data feeds.

"IOT BASED SMART FARMING STICK"

MATERIALS REQUIRED

1. ARDUINO PRO-MINI
2. 5V Power supply
3. WI-FI module
4. Sensors
4.1. Soil Moisture sensor
4.2. Temperature sensor
4.3. Smoke sensor

5. PCB plate
6. 12V battery

DESIGN PHASE OR WORKING METHOD

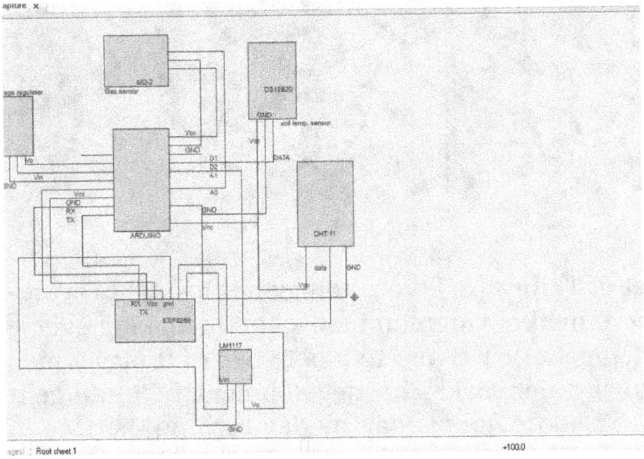

SOFTWARE REQUIRED

THINGS SPEAK AND AURDINO IDE

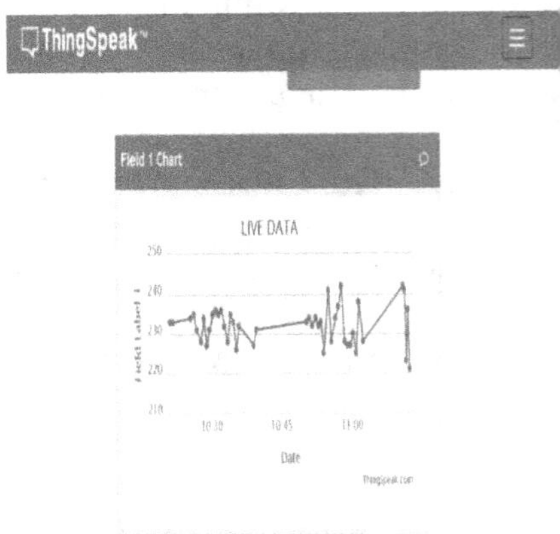

PROTOTYPE

CONCLUSION:

By proper algorithm approach in designing the controller-based system for measurement and control of several crucial parameters for growth of plants, i.e. soil moisture, humidity, light intensity, temperature and many more is under ambit. The results obtained from the evaluation after sensing of parametric values; needs to be definitive and precise. The system has to avoid quite a few many imperfections of the present systems by carefully handling the complexity, at the same time providing a flexible form of maintaining the environment. The continuously minimize costs of hardware and software, the expanded taken of electronic systems in agriculture, and growing agricultural control system industry in several region of agricultural production, will result in control systems which we can rely upon, that will provide quality and quantity of production. Improvements will be made less expensive and more reliable sensors will be developed for use in agricultural production.

FUTURE SCOPE: Future work would be focused more on increasing sensors on this project to fetch more data especially with regard to Pest Control and by also integrating GPS module in this IoT Stick to enhance this Agriculture IoT Technology to full-fledged Agriculture Precision ready product.

CASE STUDY 2: FIRE FIGHTING ROBOT

Introduction

One of the most important parameters in fire disaster is life, i.e., lives lost in saving someone else life. It is sometimes impossible for fire-fighters personnel to access the site of a fire because of explosive materials, smoke, and high temperatures. A fast response to detect the fire can avoid many disastrous things. It is observed that fire can take place at domestic as well as at industrial level. A normal spark can generate a massive fire breakout. Not only lives of industrial people but also the lives of domestic's people are at risk because of poor fire management system. Fire can take many lives and can injure many people for their lifetime. But it can be avoided using proper fire controlling methods.

For such environments, fire-fighting robot is proposed. In today's generation a lot of robots are proposed and designed to remove the human factor from dangerous and deadly work. The use of robots is becoming very common that safely completes the labour intensive or deadly work for human beings. In Fire Extinguishing robot, we intend to build a system that could extinguish a small flame by sensing and moving to the location itself.

It will automatically detect the fire with the help of flame sensors. Once it detects the fire breakout location, it navigates itself accordingly to reach the fire source and extinguishes the fire by using built-in fire extinguishing system. For fire detection it is using three flame sensors. First one for the left direction, second one for the forward direction and third one for the right direction. Fire extinguishing system will get activated when fire detection system detects fire. It then reaches the breakout point and water pump will start ejecting the water when it detects fire. The key features of this system are to provide surveillance of fire so that major fire accidents can be prevented, and loss of human lives gets minimized.

Empathizing

Fire disaster is one of the dangerous problems that can lead to heavy loss both financially and by taking lives. Sometimes it becomes difficult for fighters to access the site of a fire because of explosive materials, smoke, and high temperatures. Such situations risk the lives of fire fighters too. In such environments, fire-fighting robots can be useful. In Fire Extinguishing Robot, we intend to build a system that could extinguish a small flame by sensing and moving to the location itself. Sometime delay in the arrival of fire fighters leads to numerous consequences. The Fire Extinguishing robot continuously monitors the environment and extinguishes it without delay.

Requirements

Hardware Requirements

1. ARDUINO UNO

Arduino UNO

Figure shows the Arduino UNO board. It is basically a micro-controller kit that is used to get data from peripheral devices (sensors, motors, etc.). The Arduino UNO Micro- controller board is based on the ATmega328P IC. The ATmega328P is good platform for robotics application which makes robot to extinguish fire in real time. Arduino UNO board consists of the sets of digital and analog pins that may act as an interface to various expansion boards and other circuits. It contains everything needed to support the microcontroller.

2. Flame Sensor

Flame sensor

Figure shows the flame sensor. The flame sensor senses the environment and detects the presence of fire or flame. It detects the presence of flammable and harmful gases likes nitrogen, hydrogen, carbon mono oxide. The signal detection capacity is adjustable. The robot contains three flame sensors.

3. Servo Motor

Servo motor

Figure shows the servo motor. Servo motors are electronic devices that are mainly used for providing specific velocity and acceleration.

4. L293D Motor Driver Shield

L293D motor driver shield

Figure shows the L293D motor driver. L293D is a motor driver or motor driver IC which is responsible for the movement of DC motor on either direction. L293D is a 16-pin IC through which we are able to run two DC motors simultaneously in any direction.

5. Mini DC Submersible Pump

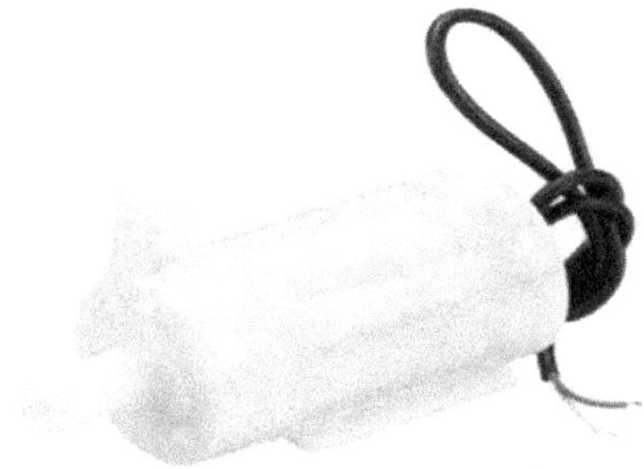

Mini DC submersible pump

Figure shows the submersible water pump. Submersible water pump is ideal for making automatic watering system using Arduino. The water pump is an important part of the robot as it will pump water to extinguish the fire.

6. Robot Chassis with Motor and Wheel

Robot chassis with motor and wheel

7. Connecting Wires

Connecting wires

A connecting wire is an electrical wire, or group of them in a cable, with a connector or pin at each end which is normally used to interconnect the components of a breadboard or other prototype or test circuit, internally or with other equipment or components, without soldering.

Software Requirements

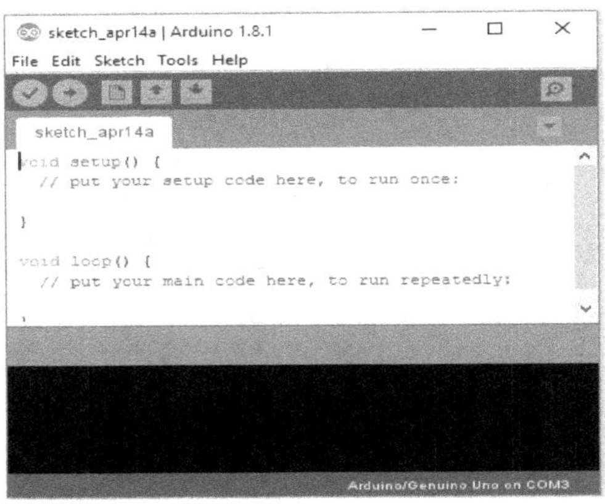

Arduino IDE

For programming, the Arduino software provides an integrated development environment (Arduino IDE) and core libraries. The Arduino IDE program is a software program written in Java language and based on the Processing. The Arduino IDE is basically a framework built on top of C and C++ and compiled using avr-gcc and AVR Libc. The open-source Arduino IDE makes it easy to write code and upload it to the Arduino Uno for execution. It is available for all major desktop platforms i.e., Windows, Mac OS X, and Linux. Figure shows the Arduino IDE program.

IDEATE

Empathy Mapping

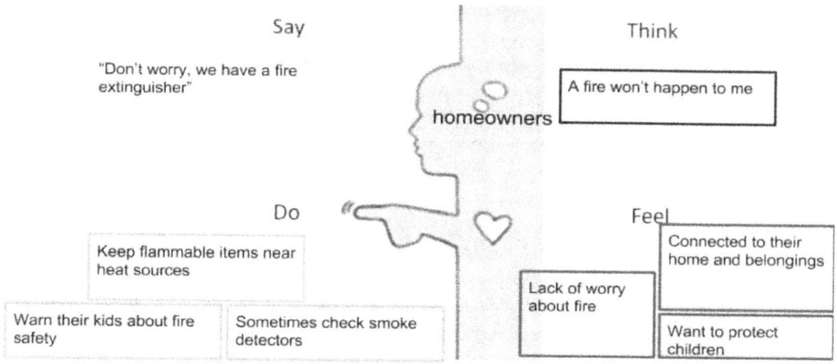

Block diagram

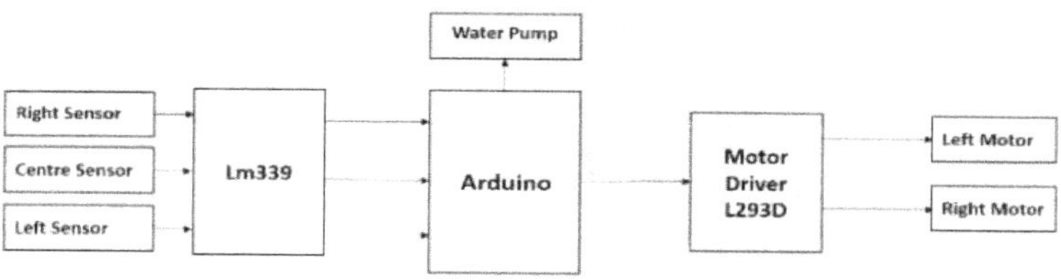

Block diagram for fire fighting robot

Circuit Diagram

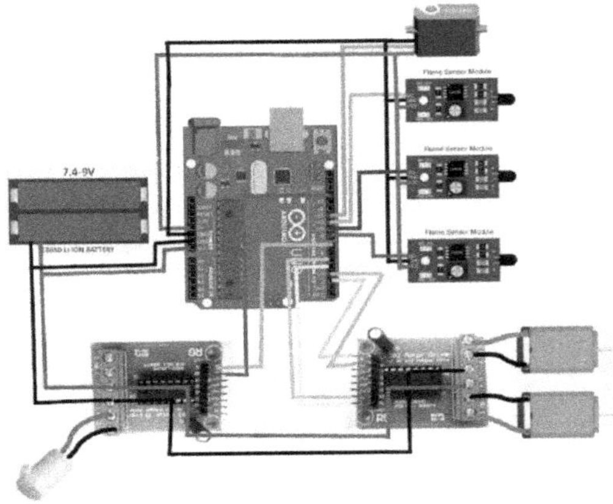

Circuit diagram for fire sensor robot

Prototyping

Prototype Evaluation

EVALUATION CRITERIA	PROTOTYPE
FUNCTIONALI TY	IT'S A DEVICE THAT WORKS AS A FIRE EXTINGUISHER, WITHOUT HUMAN SUPPORT. USING THIS THERE WILL BE LESS DAMAGE TO HUMAN KIND.
COST	RS. 2500
AESTHETICS	APPEALING STRUCTURE
USABILITY	USABLE IN DANGEROUS SITUATIONS. USED TO STOP FIRE INSTANTLY. MORE EFFICIENT.

Testing

Our fellow engineers ran through our prototype and gave us some valuable reviews:

CUSTOMER PROFILE	PROS OF CONCEPT	CONS OF CONCEPT
Name: Anushka Singh Age: 22 Student: Pursuing BTECH 3rd Year in in Graphic Area University Behaviour: Obedient, Positive Attitude	• Decreases Property Damage • Decreases Property Damage • Decreases Property Damage • Works Rapidly And Continously	• Faulty Installations May Occur • Consumes More Money • Difficult To Handle • Sensor Dependent

Conclusion

This model of Fire Extinguishing Robot aids to share out the burden of fire fighters in firefighting task. Our project aims to build a real time firefighting robot which moves in a constant speed, identify the fire and then extinguish it with the help of pumping mechanism. The detection and extinguishing were done with the help basic hardware components attached with the robot. Firstly, Flame sensors are used for the detection of fire. Secondly, Rubber wheels are used to navigate the robot to reach the fireplace. Finally, the robot extinguishes the fire with the help of submersible water pump and servo motors.

CASE STUDY 3: Unsafe Sanitation

Empathize

About 4.5 billion people (more than half of world's population) either practice open defection or use unsafe sanitation facilities and services. Globally, more than 1 billion people live in poor urban communities. Of them, nearly 300 million live without access to safe drinking water, and more than 400 million lack adequate sanitation. The majority of urban poor live in Asia (60%) and sub-Saharan Africa (22%) and reside in what the UN defines as "slums ": communities in which there is no security of tenure, no access to improved water and sanitation, where dwellings are made of non-durable materials and are generally overcrowded (more than 3 people per room).

1 billion people live in poor urban communities
1.1 billion do not have access to safe water
2.4 billion do not have access to adequate sanitation

—World Health Organization, 2000

Many slum dwellers live under all or most of these conditions. Urban populations without access to improved sanitation will grow from 661 million (in 2006) to 898 million (by 2015). The majority of these people are or will be living in slums and peri-urban settlements that are growing by over a million people per week. The number of urban residents without access to improved sanitation will increase from 661 million to 898 million by 2030. Most of these people will be living in slums and peri-urban settlements which, globally, are growing by over a million people per week.

Why Choose this Topic?

- Even though sanitation is considered a human right, many people in the developing world lack access to even basic toilets.
- Each year, 760,000 children die of sanitation-related diseases.
- In Kenya's slums, about 8 million people lack hygienic and dignified sanitation solutions, and 4 million tons of untreated sludge is dumped into waterways every year.
- People living in these informal settlements are at high risk of contracting waterborne diseases like typhoid or cholera.
- In addition to its effects on community health, poor sanitation also has implications for education, the environment, and human dignity.

Analyze

- High burden due to faecal-oral contamination
- Low coverage and scale
- Non-viable markets & delivery networks
- Expensive expert solutions
- Limited innovation capacity

4 Why Analysis

Unsafe Sanitation

↓

Increase in population growth (39.2%)

Lack of basic sanitation facilities (56%)

Insignificant inequalities in service levels (84%)

Least development countries in world (16.8%)

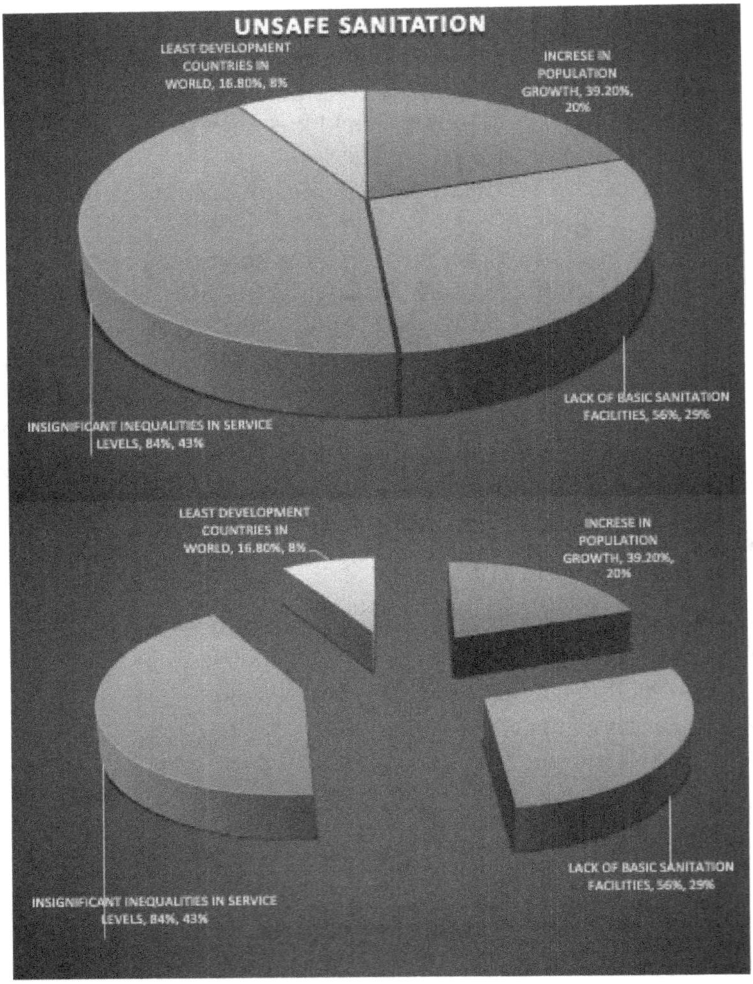

Potential Causes

- Insignificant inequalities in service levels
- lack of basic sanitation facilities

Implications

- Demand for housing, services, and infrastructure.
- Without tenure often unable to connect to services.
- Risks of water insecurity and inequitable services.

Integrated Approaches

- Across all areas
- Linking services, planning and governance
- Cities generate 40 billion litres of sewage daily (100L/capita). 10 Billion is "treated" (25%).

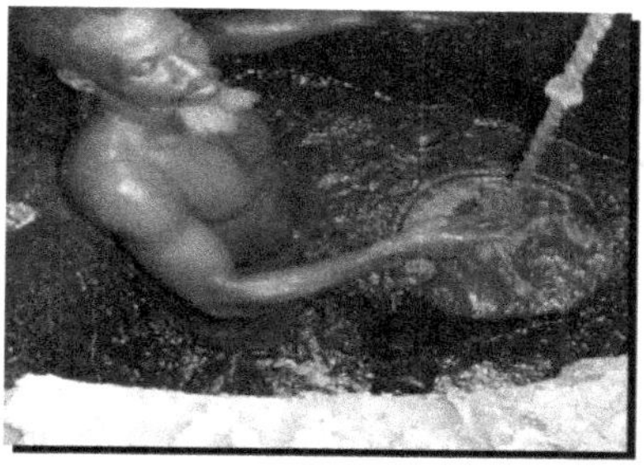

Process

Human Centered Design: Understanding people's needs and desires to identify new solutions that will be desirable, feasible, and viable.

What if?

- The sanitation value chain could be made more efficient with innovation.
- Expanding and improving sanitation without central sewers, because this is one and will be one of the most common type of sanitation service used by the poor.
- Making sanitation services safe and sustainable by addressing the failure to effectively transport, treat, and reuse waste captured in on-site facilities.

Addressing Problems

Families living in shared compounds

Most people in Kumasi live in rented accommodations, usually sharing a room with their family in a compound of 9 or 10 such rooms and a common area, so we are designing a solution for the majority of people.

Some level of family income, even if inconsistent

Most family members in the poor communities are unemployed so we are targeting families with some amount of discretionary income who can pay for monthly sanitation services.

Currently using public toilets

For people who do not have an in-home toilet, paid public toilets or open defecation are the only options. We are focusing on providing an improved sanitation solution for those already paying daily for defecation.

Desire in-home sanitation, but not able to afford existing options

For most, the cost of a $700 pit latrine is not a viable option financially so we are offering less expensive alternatives.

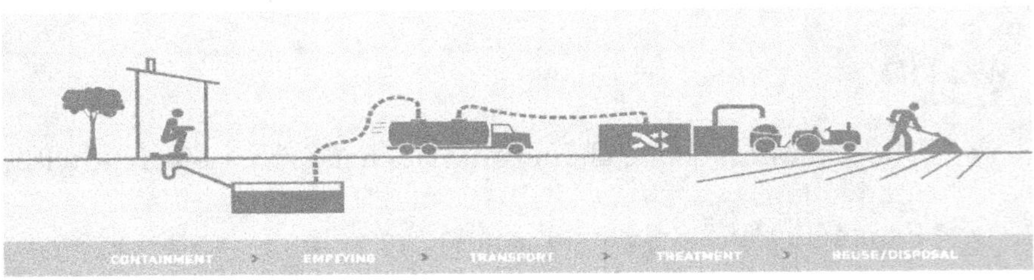

Our Insights

- Public toilets have become the primary toilets.
- The nearest public toilet is sometimes a 15-minute walk from home.
- There are long lines for the toilets every morning and evening.
- Public toilets are especially difficult for the elderly and sick to use.
- Many families still keep a chamber pot in the home, and then dump it.

In-home sanitation services are nothing new, but many are now illegal.

- Bucket latrines were outlawed because service collectors dumped waste illegally, rather than in designated collection locations.
- Though illegal, bucket latrines continue to be used.
- Night soil collection was (and still is in some areas) a formal business with contracts, scheduled pickup dates, and shared payments.
- Some homes still have an empty room where bucket latrines used to be installed.

Pee and poo are different.

- People are willing to pay to use the public toilets for pooping, but not for peeing.
- Men often pee on the side of the road or against a wall, and its accepted behaviour.
- Women pee into a chamber pot or into a drain in the floor of the bathroom.
- It's common for children to poop outside even though it's taboo for adults.

The compound is the home, and it's almost always rented.

- Each family lives in a room surrounding the common area. Most rooms are very small, and many people sleep on the floor.

- There are defined common areas within the compound for certain activities such as bathing and cooking.
- Chamber pots are kept in the bedroom or bathroom.
- People living in the compound share certain expenses such as water and electricity.
- Upgrades and improvements to the compound are difficult because consensus is difficult to achieve.
- Rent is often paid for 1-4 years upfront, so landlords have minimal incentive to upgrade or improve.

Ideate and Solve

- We can provide toilets which can purify the waste produced by themselves and later purify it, as well as use it for any purpose.
- We can support initiatives that help stimulate market and community demand for improved sanitation in order to accelerate development of productive sanitation market conditions.
- We can work with local governments, service providers, and community-based organizations in select places to foster environments that support the use of community-responsive, non-severed sanitation products, delivery methods, and business models.
- Providing city-wide, inclusive sanitation requires new service models and market structures, as well as different approaches and technologies to serve different community needs and settings.
- We can educate people about the diseases caused by it and create camps to help them in many ways.

Reinventing the Toilet

Using transformative technologies throughout the sanitation value chain to plug key gaps.

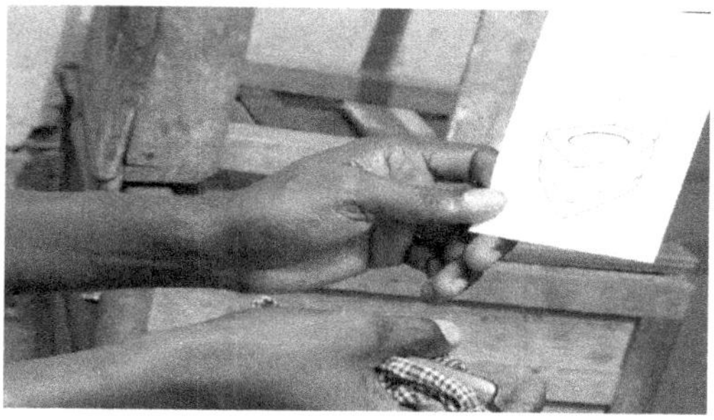

Reinvented Toilet

Reinvented Toilet

- No pathogens
- User demand
- 5 kg per day
- Less than 25% total solids

- Access 90% of pits

Omni- processor

- Access 90% of pits
- Effective emptying
- Effective transport
- 50 GPM
- 0 to 40% solids

- Access 90% of pits

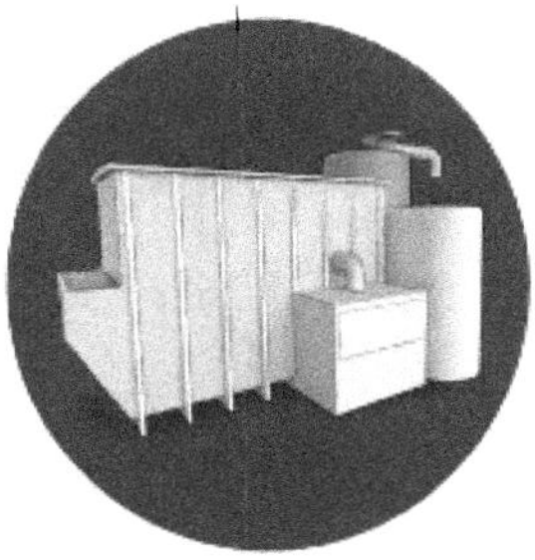

Omni-ingestor

- Decentralized
- Energy production
- Nutrient recovery
- 20 to 50% of solids
- 20 tons per day

The Omni Processor: Turning Sewage into Drinking Water

Taking the idea of one man's trash being another man's treasure to new lengths, the "omni processor" plant takes literal human waste and turns it into valuable goods - including electricity and clean drinking water.

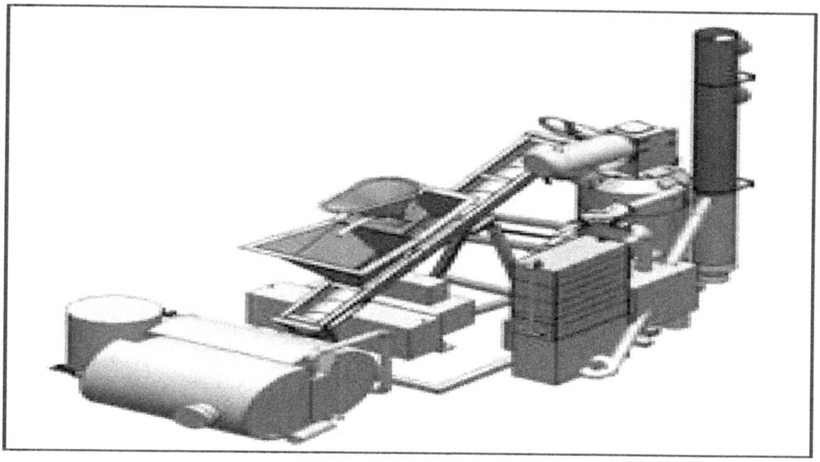

How Does it Work?

- The Omni Processor works a little bit like a steam power plant, an incinerator, and a water filtration system combined - with all three of them working in harmony with one another.
- The wet sewage enters the plant first and is boiled to 100 degrees and subsequently separated into dry solids and water vapor.
- The incinerator then burns those dry solids, producing a high-pressure steam.
- That steam helps to drive a steam engine, which via a generator produces the electricity which powers the Omni Processor.

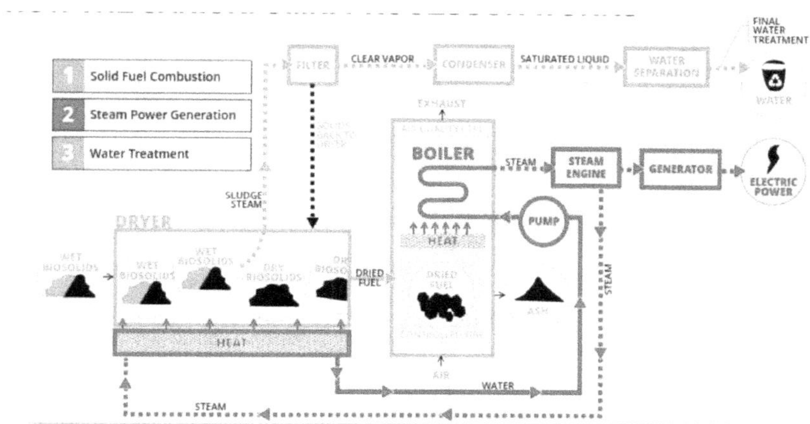

- That steam is then condensed back into water and is filtered multiple times, turning it into pure, distilled water, suitable for drinking.
- The electricity generated is enough to power the cycle of processes, meaning the Omni Processor is self-sustaining. And as if that wasn't enough, the plant also generates additional electricity which can be distributed elsewhere, and the solids left by the boiler - a kind of non-toxic ash-contain phosphorus and potassium, which can be used as a fertilizer for soil.

The Product/Service Cost Trade-off

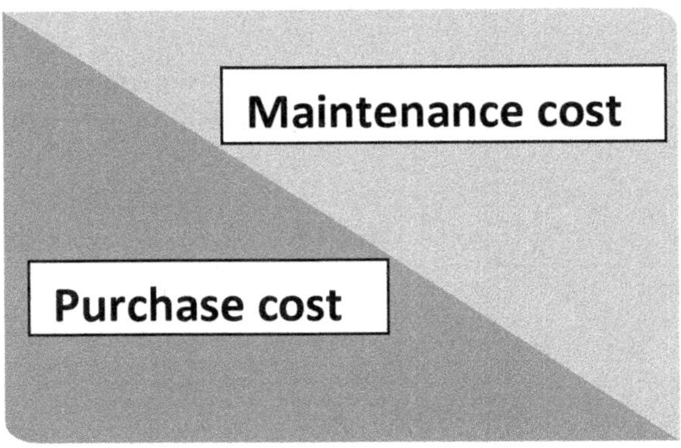

- To design a toilet with low up-front costs, a trade-off must be made in terms of capacity.
- This results in higher ongoing service costs.
- A consumer can save on maintenance costs by investing more upfront, but this is rarely feasible financially.

Our Opportunities

1. High-Touch Service Toilet

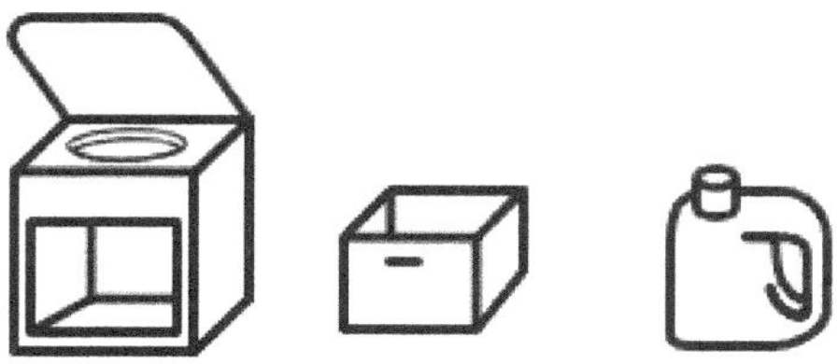

This is a sanitation service where users lease a toilet with interchangeable waste cartridges that are emptied and serviced on a bi-weekly basis.

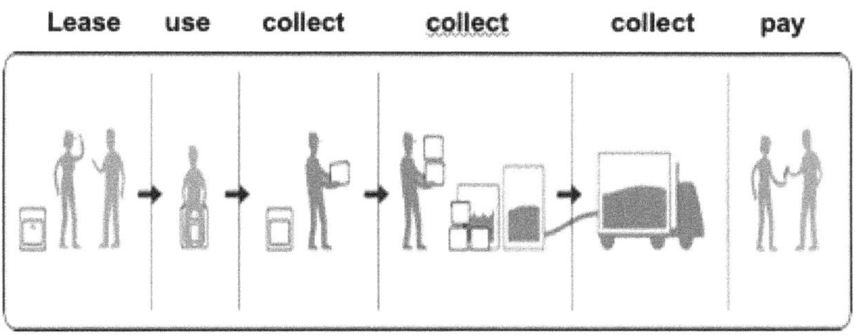

Buy

- The customer signs up for the service which includes a deposit from a door-to-door salesperson.
- The salesperson demonstrates use of the toilet, schedules service and collects the first month's payment.
- Door-to-door sales & service set-up

Use

- The customer places the toilet in either the bathroom (or common area with a privacy shelter) as appropriate.
- No special maintenance by the user

Collect

- A vacuum truck service picks up the waste and brings it to the Dump site.
- Service exchange cartridge
- Cartridge is emptied and cleaned at a service station
- Vacuum truck takes waste to treatment site

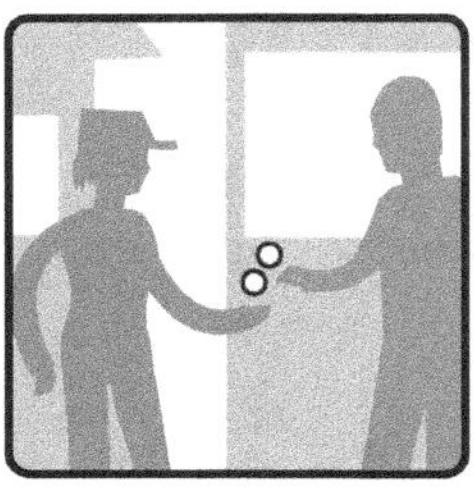

Pay

- The door-to-door salesperson returns monthly to collect the subsequent month's payment
- Salesperson returns monthly for payment collection

Benefits

- Dumping is prevented by tamper-proof design
- No plastic wastes
- Good isolation of waste during transport
- Low upfront payment for the customer
- No setup required and small hardware can fit in the bathroom or a small compound area.

Concerns

- Limited capacity depends heavily on reliability of service
- New infrastructure is required in building and outfitting a service station
- Critical mass adoption is needed in each neighbourhood to succeed.

Questions for prototyping

- How can we ensure service providers are doing their job well?
- What's the ideal product size, holding capacity, and service frequency?
- How can we recruit and incentivize franchisees?
- What's the best payment method and schedule?
- What are customer expectations around in-home service?

2. Bag and Bin

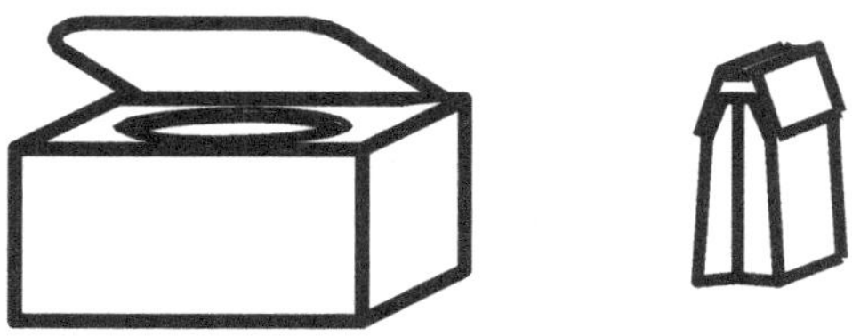

This is a low-cost system created from disposable bags and a simple toilet frame. Once a bag is full, the customer removes the bag from the toilet and moves it to a garbage storage container outside of the home. A weekly service collects the full bags.

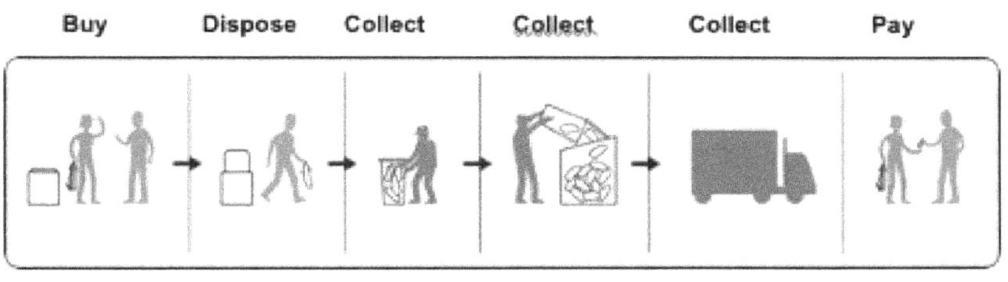

Buy

- The customer signs up for the service which includes the purchase of a toilet, a collection bin, toilet liners, and the first month of service from a door-to-door salesperson.
- The salesperson demonstrates use, installs and secures the collection bin, schedules the first weekly pickup, and collects payment.
- Door-to-door sales & set-up

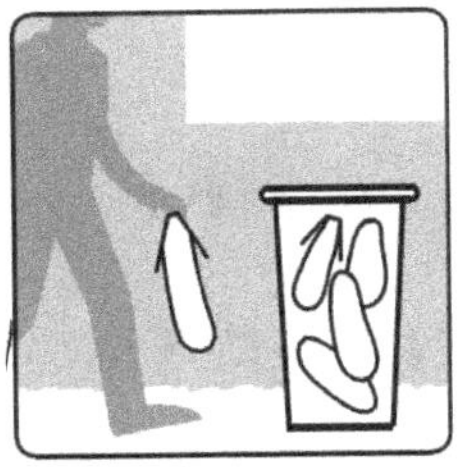

Use

- After approximately 15 uses, the bag seals and the customer carry it to the collection bin in front of the home.

Dispose

- The waste collector removes the collection bin and places it on the collection cart.
- The collector services additional households until his cart is full. He then heads to the dumpster.
- Alternatively, the collection bin is lined with a bag which is collected, and the bin is left in place.

Collect

- The collector opens the dumpster, which is locked to contain the faecal matter and prevent other waste and empties the bags from the bins which he then returns to the households.
- Weekly collection/ Waste stored in central spot/ Dump truck brings waste to treatment site.

Pay

- Every month, the salesperson collects payment and supplies the household with liners.
- Salesperson returns for monthly payment & restocks liners.

Benefits

- No setup required
- Individual or family purchase
- Low upfront payment for the customer
- Integrated hardware and service
- Waste is contained and sealed

Concerns

- New distribution network
- Optimized by large scale participation
- Large capital expenditure to launch business
- Potentially detrimental environmental impact
- Limited toilet capacity depends heavily on service
- Logistically challenging to implement

Questions for prototyping

- What's the ideal product size and holding capacity/service frequency?
- How can we ensure proper use and disposal of liners by customers?
- How can we ensure service providers are doing their jobs well?
- How can we properly incentivize service providers?
- What is the best payment method and schedule?

What we need to do

- Align local partners for on the ground implementations and day to day operations
- Get the support of local leaders
- Select target areas for prototyping
- Create a variety of toilet prototypes
- Design service models to prototype
- Build a more robust business model
- Secure philanthropic support for a pilot

Test

Conducting trails by providing them mobile toilets, which are inexpensive, as well as cost free and create awareness about proper use of sanitation.

Observations

1. They practice safe sanitation by using mobile toilets.
2. They don't have health issues, and they are aware of it.
3. They reuse purified water for commercial purposes.
4. They are aware of insignificant inequalities in service levels.
5. They provide safe, hygienic, convenient, healthy and protected sanitation for all in difficult situations.

Conclusion

Based on trials conducted, we can suggest the solution of the problem of unsafe sanitation will be following:

1. Providing mobile toilets (cost free).
2. Creating awareness about unsafe sanitation.
3. Providing safe, less infectant and hygienic toilets.
4. Protesting against insignificant inequalities in service levels.

Case Study 4: Smart City Technologies

1. Introduction

Now a days with the growing population and energy demand we should take a renewable option of energy source and also, we should keep in mind that energy should not cause pollution and other natural hazards. In this case the solar energy is the best option for us. India is a highly populated country, so we should take the advantage of such an energy which requires a very less space to produce energy efficiently.

Smart cities use a variety of software, user interfaces and communication networks alongside the Internet of Things (IoT) to deliver connected solutions for the public. Of these, the IoT is the most important. The IoT is a network of connected devices that communicate and exchange data. This can include anything from vehicles to home appliances and on-street sensors. Data collected from these devices is stored in the cloud or on servers to allow for improvements to be made to both public and private sector efficiencies and deliver economic benefits and improvements to the lives of citizens.

Many of the IoT devices use edge computing, which ensures that only the most relevant and important data is delivered over the communication network. In addition, a security system is implemented to protect, monitor and control the transmission of data from the smart city network and prevent unauthorized access to the IoT network of city's data platform.

Empathy

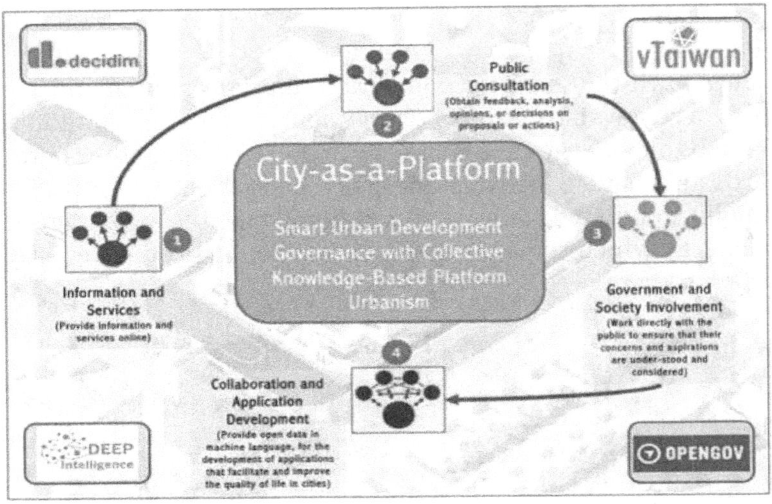

Balancing Technological Innovation with Human Values

Empathy acts as a compass, guiding the ethical and responsible use of technology. As smart cities harness the power of data and automation, it is crucial to ensure that these advancements align with human values, privacy concerns, and ethical considerations. Striking a balance between technological innovation and human- centric values helps build trust among residents, ensuring that the benefits of smart cities are realized without compromising individual rights.

Resilience and Well-being

Empathetic smart cities prioritize the well-being and resilience of their inhabitants. By understanding the daily challenges faced by residents, such as traffic congestion, pollution, and stress, city planners can deploy smart solutions that alleviate these issues and contribute to an overall improved quality of life. Integrating technologies that enhance mental health, promote sustainability, and create vibrant communal spaces reflects a commitment to the holistic well-being of the population.

Community Engagement

An empathetic smart city actively engages its residents in the decision-making process. Community input and feedback become valuable resources in shaping urban policies and technology implementations. Town hall meetings, surveys, and digital platforms can be leveraged to create a collaborative environment where citizens feel heard and valued, fostering a sense of ownership and pride in their city.

Inclusive Design

Empathy is key to crafting inclusive smart cities that cater to the needs of every individual, irrespective of age, ability, or background. Incorporating universal design principles ensures that smart infrastructure and services are accessible to all residents. For instance, smart public spaces can be designed with features that accommodate the differently abled, promoting a sense of inclusivity and belonging.

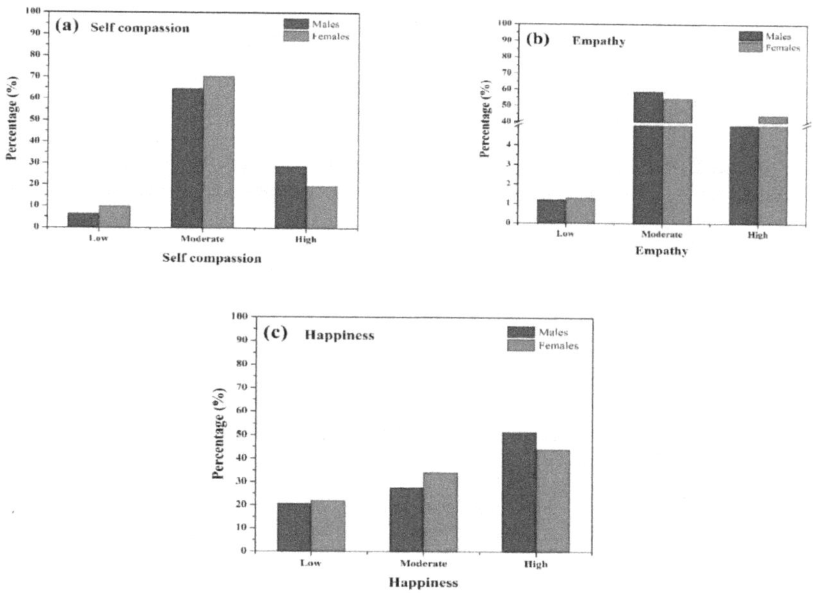

Define

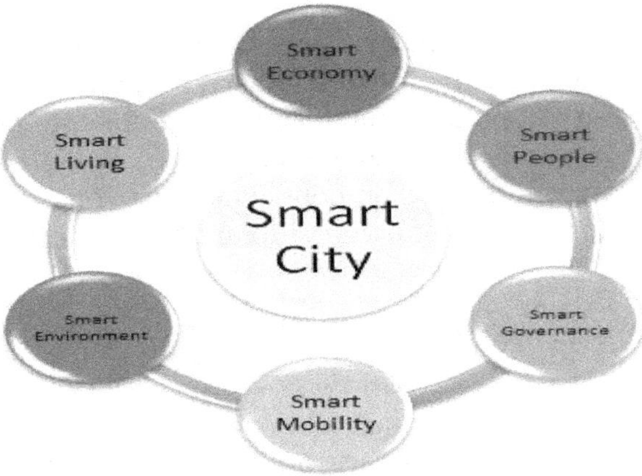

Smart cities for people refer to urban environments that leverage advanced technologies and data-driven solutions to enhance the quality of life, well-being, and overall experiences of their residents. These cities prioritize human-centric approaches in the design and implementation of innovative solutions, placing the needs, preferences, and inclusivity of the population at the forefront of urban development.

Human-Centric Design

The development of smart cities is guided by an understanding of the diverse needs, behavior, and aspirations of the people who live in them. Urban planners and policymakers prioritize creating environments that cater to the well-being and satisfaction of residents.

Inclusivity

Smart cities for people embrace inclusivity by ensuring that technological solutions are accessible and beneficial to all members of the community, regardless of age, ability, socioeconomic status, or cultural background. Inclusive design principles are applied to infrastructure, public spaces, and services.

Community Engagement

Residents actively participate in decision-making processes through various channels such as community forums, surveys, and digital platforms. This engagement fosters a sense of shared responsibility, ownership, and empowerment among the city's inhabitants.

Technological Integration

Smart cities leverage cutting-edge technologies, including the Internet of Things (IoT), artificial intelligence (AI), data analytics, and sustainable energy solutions. These technologies are seamlessly integrated into the urban fabric to improve efficiency, resource management, and overall urban functionality.

Sustainability

There is a strong emphasis on environmental sustainability, resource efficiency, and the reduction of the city's ecological footprint. Smart cities for people prioritize eco- friendly initiatives, such as renewable energy, waste reduction, and green spaces, contributing to a healthier and more sustainable urban environment.

Resilience

Smart cities are designed to be resilient in the face of challenges such as natural disasters, economic fluctuations, and social disruptions. Robust infrastructure, advanced emergency response systems, and adaptable urban planning contribute to the city's ability to recover and thrive.

Privacy and Ethics

The ethical use of technology and the protection of individual privacy are paramount in smart cities for people. Stringent data protection measures, transparency in data usage, and responsible AI deployment ensure that technological advancements benefit residents without compromising their rights.

Ideate

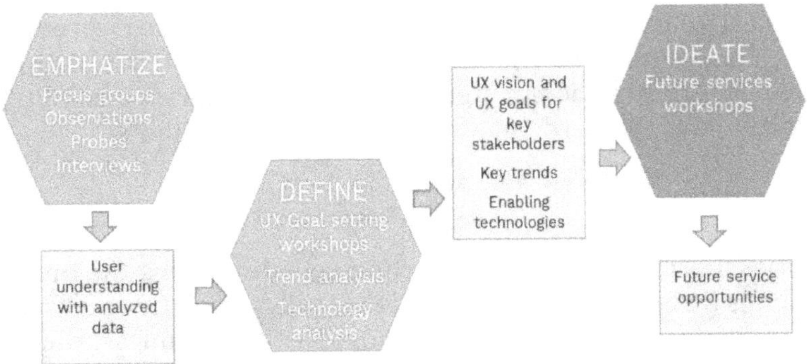

Interactive Public Spaces

- Implement interactive public spaces equipped with smart benches, interactive art installations, and augmented reality zones, fostering community engagement.
- Integrate smart lighting and seating arrangements that adapt to the needs and preferences of individuals, promoting comfort and inclusivity.

Community-Driven Urban Planning

- Establish digital platforms for community input in urban planning decisions, allowing residents to propose and vote on initiatives that directly impact their neighborhood.
- Use virtual reality (VR) simulations to involve citizens in the design process, providing a visual representation of proposed changes and gathering feedback.

Smart Mobility Solutions

- Develop an integrated and sustainable public transportation system with real- time tracking, optimizing routes based on demand and traffic conditions.
- Implement smart parking solutions using sensors and mobile apps to guide drivers to available parking spaces, reducing traffic congestion.

Smart Education Hubs

- Establish smart education hubs with high-tech classrooms, interactive learning tools, and virtual collaboration spaces to facilitate innovative and inclusive learning experiences.
- Implement smart libraries that provide digital access to educational resources and offer community programs for continuous learning.

Digital Citizenship Programs

- Launch educational programs to enhance digital literacy and empower residents to make informed use of smart city technologies.
- Encourage active participation in local governance through digital platforms, fostering a sense of responsibility and community involvement.

Safety and Security Solutions

- Deploy smart surveillance systems with AI capabilities to enhance public safety while respecting privacy.
- Develop community-based neighborhood watch applications that facilitate communication and collaboration among residents to improve overall safety.

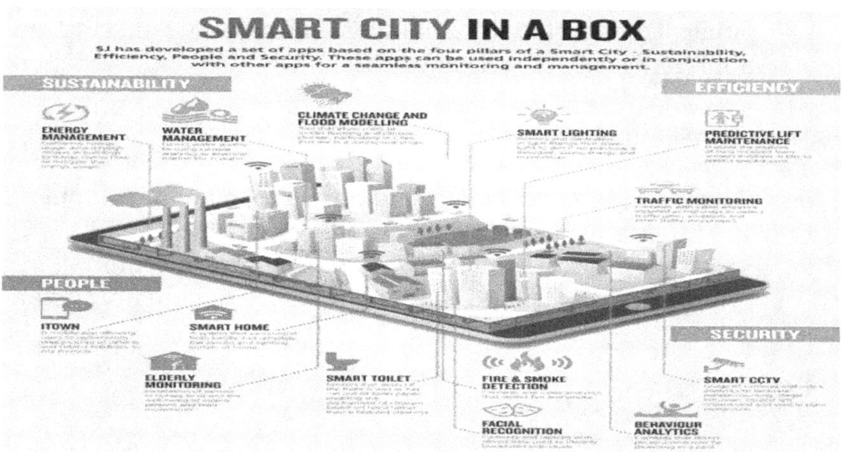

Prototype

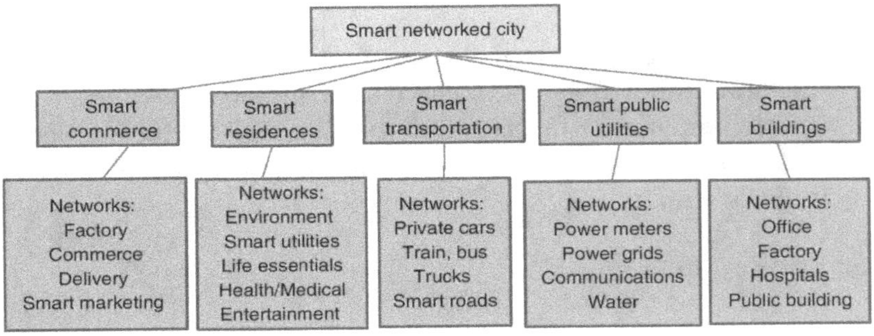

Interactive Community Hub

- Hardware: A physical hub in a central location equipped with interactive touchscreens, augmented reality displays, and community bulletin boards.
- Software: An integrated platform that allows residents to access information about local events, services, and engage in community discussions.

Smart Mobility Hub

- Hardware: Smart kiosks equipped with touchscreens and sensors at major transportation hubs.
- Software: An application that provides real-time updates on public transportation schedules, nearby ride-sharing options, and available parking spaces.

Digital Citizenship Initiatives

- Hardware: Digital literacy centers equipped with computers and interactive learning materials.
- Software: Educational apps and programs promoting responsible use of technology, online safety, and digital citizenship.

Interactive Learning Zones

- Hardware: Smart classrooms with interactive displays, VR headsets, and collaborative workspaces.
- Software: Educational apps that supplement traditional learning methods, encouraging interactive and personalized learning experiences.

Integration and User Experience

- Centralized App: Residents can access all smart city services through a centralized mobile application, providing a seamless and user-friendly experience.
- User Feedback Mechanism: An integrated feedback system allows residents to provide feedback on various services and suggest improvements.
- Community Engagement Events: Regular community events hosted at the Empathy Hub to foster interaction, collaboration, and a sense of belonging.

This prototype concept envisions a comprehensive approach to creating a smart city that prioritizes people's needs, enhances community engagement, and promotes sustainability and inclusivity. The integration of hardware and software components aims to create a holistic and user-centric urban living experience.

Test

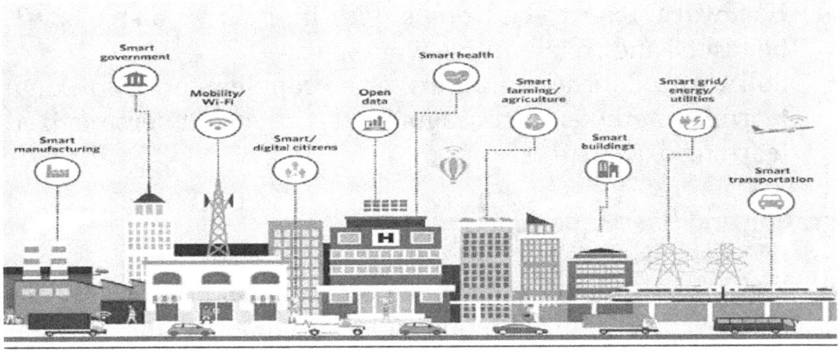

Objective

To evaluate the effectiveness and user satisfaction of the implemented smart city features with a focus on enhancing the quality of life for residents.

Usability Testing

- Evaluate the ease of use of digital interfaces, such as mobile apps, interactive kiosks, and community display boards.
- Test the accessibility of information for users with different abilities, including those with visual or auditory impairments.
- Assess the efficiency of navigation and task completion within the smart city applications.

Inclusivity Assessment

- Analyze the inclusivity of public spaces by assessing the comfort and accessibility for diverse demographic groups, including the elderly, differently- abled individuals, and various cultural backgrounds.
- Verify the effectiveness of inclusive design elements in public areas, such as adaptive seating and sensory-friendly features.

Sustainability and Resilience Testing

- Evaluate the efficiency of green infrastructure elements, such as energy-efficient buildings and waste management systems.
- Test the responsiveness and effectiveness of the early warning system for natural disasters.

Learning and Educational Impact

- Measure the engagement levels and effectiveness of interactive learning zones, considering factors like increased knowledge and skill acquisition.
- Assess the usage and impact of digital literacy centers on the community's overall digital literacy.

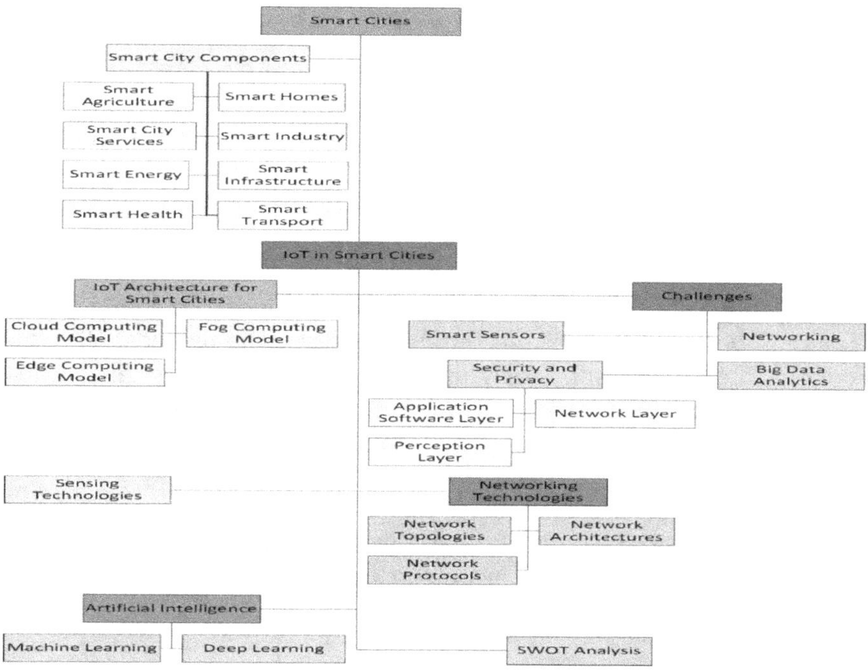

Conclusion

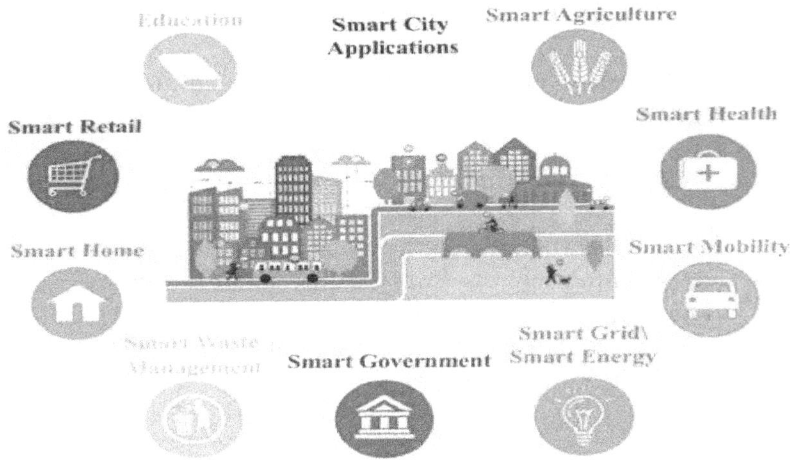

Creating smart connected systems for our urban areas provides a great many benefits for citizens around the world, not only to improve quality of life, but also to ensure sustainability and the best possible use of resources.

These solutions are dependent on a unified approach from government as well as the private sector and residents themselves. With the correct support and infrastructure, however, smart cities can use advances such as the Internet of Things to enhance the lives of residents and create joined up living solutions for the growing global urban citizenry.

TWI has expertise in many of the underpinning technologies of a new smart future, including sensors, electrification and power generation solutions.

www.ingramcontent.com/pod-product-compliance
Lightning Source LLC
LaVergne TN
LVHW052324200926
848481LV00002B/666
9781922617767